USES OF RISK ANALYSIS TO ACHIEVE BALANCED SAFETY IN BUILDING DESIGN AND OPERATIONS

Committee on Risk Appraisal in the Development of Facilities Design Criteria
Building Research Board
Commission on Engineering and Technical Systems
National Research Council

Bruce D. McDowell
Andrew C. Lemer
Editors

NATIONAL ACADEMY PRESS
Washington, D.C. 1991

NOTICE: The project that is the subject of this report was approved by the Governing Board of the National Research Council, whose members are drawn from the councils of the National Academy of Sciences, the National Academy of Engineering, and the Institute of Medicine. The members of the committee responsible for the report were chosen for their special competences and with regard for appropriate balance.

This report has been reviewed by a group other than the authors according to procedures approved by a Report Review Committee consisting of members of the National Academy of Sciences, the National Academy of Engineering, and the Institute of Medicine.

The National Academy of Sciences is a private, nonprofit, self-perpetuating society of distinguished scholars engaged in scientific and engineering research, dedicated to the furtherance of science and technology and to their use for the general welfare. Upon the authority of the charter granted to it by the Congress in 1863, the Academy has a mandate that requires it to advise the federal government on scientific and technical matters. Dr. Frank Press is president of the National Academy of Sciences.

The National Academy of Engineering was established in 1964, under the charter of the National Academy of Sciences, as a parallel organization of outstanding engineers. It is autonomous in its administration and in the selection of its members, sharing with the National Academy of Sciences the responsibility for advising the federal government. The National Academy of Engineering also sponsors engineering programs aimed at meeting national needs, encourages education and research, and recognizes the superior achievements of engineers. Dr. Robert M. White is president of the National Academy of Engineering.

The Institute of Medicine was established in 1970 by the National Academy of Sciences to secure the services of eminent members of appropriate professions in the examination of policy matters pertaining to the health of the public. The Institute acts under the responsibility given to the National Academy of Sciences by its congressional charter to be an adviser to the federal government, and upon its own initiative, to identify issues of medical care, research, and education. Dr. Stuart Bondurant, is acting president of the Institute of Medicine.

The National Research Council was established by the National Academy of Sciences in 1916 to associate the broad community of science and technology with the Academy's purposes of furthering knowledge and of advising the federal government. Functioning in accordance with general policies determined by the Academy, the Council has become the principal operating agency of both the National Academy of Sciences and the National Academy of Engineering in providing services to the government, the public, and the scientific and engineering communities. The Council is administered jointly by both Academies and the Institute of Medicine. Dr. Frank Press and Dr. Robert M. White are chairman and vice chairman, respectively, of the National Research Council.

Funding for the FCC program was provided through the following agreements between the indicated federal agency and the National Academy of Sciences: Department of State Contract No. 1030-621218; National Science Foundation Grant No. MSS-8902669, under master agreement 8618641; and U.S. Postal Service grant, unnumbered.

Library of Congress Catalog Card No. 91-67599
International Standard Book Number 0-309-04680-7

Additional copies of this report are available from:

National Academy Press
2101 Constitution Avenue, NW
Washington, DC 20418

S 493

Printed in the United States of America

COMMITTEE ON RISK APPRAISAL IN THE DEVELOPMENT OF FACILITIES DESIGN CRITERIA

Chairman

BRUCE D. McDOWELL, Director Government Policy Research, Advisory Commission on Intergovernmental Relations, Washington, D.C.

Members

JOHN C. CANESTRO, P.E., Consultant, Pleasanton, California
MICHELLE A. DEPEW, Building Inspector, Brewster, New York
BRUCE ELLINGWOOD, Professor and Chairman of Civil Engineering, The Johns Hopkins University, Baltimore, Maryland
ROBERT WILLIAM FITZGERALD, Professor of Civil Engineering, Worcester Polytechnic Institute, Worcester, Massachusetts
GEORGE L. HEAD, Vice President, Underwriters and Insurance Institute of America, Malvern, Pennsylvania
FREDERICK KRIMGOLD, Associate Dean for Research and Extension, College of Architecture and Urban Studies, Virginia Polytechnic Institute, Washington-Alexandria Center, Alexandria, Virginia
ANATOL LONGINOW, Wiss, Janney, Elstner Associates, Inc., Northbrook, Illinois
MORRIS A. WARD, Executive Director of Environmental Health Center, National Safety Council, Washington, D.C.
CHRIS WHIPPLE, Vice President and Director, Western Operations, Clement International Corporation, San Francisco, California

ROBERT G. ZALOSH, Manager of Applied Research Department and Assistant Vice President, Factory Mutual Research Corporation, Norwood, Massachusetts

Federal Construction Council Liaison Representatives

RICHARD BUKOWSKI, Fire Hazard Analysis, National Institutes of Standards and Technology, Gaithersburg, Maryland
ROBERT CARR, Architectural Service, Office of Facilities, Department of Veterans Affairs, Washington, D.C.
JEFF CISNEY, Department of Veterans Affairs, Washington, D.C.
JOSEPH J. CORLISS, Division of Health Facilities Planning, HHS/PHS, Parklawn Building, Rockville, Maryland
DONALD EVICK, Real Estate and Building Department, U.S. Postal Service, Washington, D.C.
DAVID FRABLE, General Services Administration, Washington, D.C.
TOM HURLEY, Design Policy Branch, Naval Facilities Engineering Command,Alexandria, Virginia
ROBERT T. KRATZKE, U.S. Department of Energy, Washington, D.C.
RICH LAMPO, U.S. Army Construction Engineering Research Laboratory, Champaign, IL
JACK METZLER, U.S. Department of Energy, Washington, D.C.
JUANITA MILDENBERG, Facilities Engineering Branch, National Institutes of Health Bethesda, Maryland
KENNETH MURPHY, U.S. Department of Energy, Germantown, Maryland
CHARLES B. PITTINGER, JR., P.E., Facilities Engineering Office, National Aeronautics & Space Administration, Washington, D.C.
STROHN SPARKS, Office of Health Facilities, HHS/PHS, Rockville, Maryland
WAYNE STIEFEL, CFR, National Institute of Standards and Technology, Gaithersburg, Maryland

Public Facilities Council Liaison Representatives

ROBERT KELLER, Division of Design and Construction, Office of Administration, State of Missouri, Jefferson City, Missouri

Building Research Board

ANDREW C. LEMER, Director
HENRY A. BORGER, Executive Secretary, FCC
PATRICIA WHOLEY, Staff Associate
SUZETTE CODY, Project Assistant
JOANN CURRY, Senior Secretary

BUILDING RESEARCH BOARD
(1990-1991)

Chairman

RICHARD T. BAUM, Consultant, Jaros, Baum and Bolles, New York, New York

Members

LYNN S. BEEDLE, University Distinguished Professor of Civil Engineering and Director, Council on Tall Buildings and Urban Habitat, Lehigh University, Bethlehem, Pennsylvania
GERALD L. CARLISLE, Secretary-Treasurer, International Union of Bricklayers & Allied Craftsmen, Washington, D.C.
NANCY RUTLEDGE CONNERY, Consultant, Woolwich, Maine
C. CHRISTOPHER DEGENHARDT, President, EDAW, Inc., San Francisco, California
ELISHA C. FREEDMAN, Regional Manager, Boyer, Bennett & Shaw, Inc., and Executive-in-Residence, University of Hartford, Connecticut
DONALD G. ISELIN, USN, Retired, Consultant, Santa Barbara, California
FREDERICK KRIMGOLD, Associate Dean for Research and Extension, Virginia Polytechnic Institute and State University, Alexandria, Virginia
J. W. MORRIS, USA, Retired, Engineer Advisor, Zorc, Rissetto, Weaver & Rosen, Washington, D.C.

HAROLD J. PARMELEE, President, Turner Construction Company, New York, New York

LESLIE E. ROBERTSON, Director, Design and Construction, Leslie E. Robertson Associates, New York, New York

JAMES E. WOODS, William E. Jamerson Professor of Building Construction, College of Architecture and Urban Studies, Virginia Polytechnic Institute and State University, Blacksburg, Virginia

APRIL L. YOUNG, Senior Vice President, NVR Development, L.P., McLean, Virginia

Staff

ANDREW C. LEMER, Director
HENRY A. BORGER, Executive Secretary, Federal Construction Council
PETER H. SMEALLIE, Executive Secretary, Public Facilities Council
PATRICIA M. WHOLEY, Staff Associate
SUZETTE R. CODY, Project Assistant
JOANN V. CURRY, Senior Secretary
LENA B. GRAYSON, Program Assistant

PREFACE

A chain is only as strong as its weakest link. If one wishes to carry a heavy burden, the strength of all links in the chain must be at least adequate for the load, and there is little value in having a few stronger links if weak ones break. And so it is with safety of buildings.

Our current system of building codes, design criteria, and regulatory agency practices for assuring safety seems to work well for some hazards but fails to address others, and lacks mechanisms for balancing efforts to address many sources of potential risk. Communities often move quickly to alter the fire safety standards following deadly fires but may neglect for years the threat of earthquakes. As a nation we spend millions of dollars to remove asbestos from older buildings while remaining relatively indifferent to radon gas.

Federal government agencies are entrusted with using the public's resources as efficiently as possible to achieve their missions, and must respond to this inconsistent attention to safety when legislation or other regulations mandate uneven or uninformed strategies for achieving safety. Some of these agencies, drawing on the experience of managing defense and nuclear systems and regulating food and drug safety, urge broader application of the principles and procedures of risk analysis to the field of building safety to enhance safety cost-effectively.

The agencies of the Federal Construction Council asked the Building Research Board to consider this proposition. That request motivated the study reported here. We and the committee whose deliberations are reported here recognize that life is full of risks, and we have only limited resources to devote to building safety. As a nation, we must use all available tools to assure the safest possible buildings within the limits of our resources. We believe that risk analysis tools will enable government policy makers and building professionals to do this job better. We hope through this report to foster use and further development of these tools, and thereby to enhance the safety of America's buildings.

Bruce D. McDowell, *Chairman*
Committee on Risk Appraisal in the Development of Facilities Design Criteria

Andrew C. Lemer, *Director*
Building Research Board

EXECUTIVE SUMMARY

Risks—in buildings,[1] as elsewhere—are an unavoidable part of life. The sources of risk sometimes lead to catastrophes. The famous Chicago fire of 1871, the 1906 and 1989 San Francisco earthquakes, Boston's 1942 Coconut Grove fire, and New York's turn-of-the-century dark, stagnant, disease-ridden tenements illustrate vividly some of the more substantial dangers that people may face in buildings and other facilities.

Buildings and other facilities are expected to house and serve a variety of activities with a high degree of safety and security, and the task of ensuring that they do so has been entrusted primarily to locally enforced building codes. These codes embody criteria of acceptable design and construction practice that have developed over a period of many years, sometimes in reaction to public health and safety catastrophes resulting from growing concentrations of urban population and the introduction of dangerous new technologies. Federal government agencies have adopted their own design criteria, which are often similar to those contained in codes. By documenting for designers and facilities operators the standards of good practice defined by industry consensus, building codes and formal design criteria have made great strides in bringing facilities dangers under control.

---

[1]Much of the discussion throughout this report applies equally to all types of constructed facilities, and not only to buildings.

However, the scope of building codes and design criteria are necessarily limited. As regulatory devices operating within a local political context, codes deal with specific hazards and sometimes focus greater attention on some hazards than on others. While some code provisions and criteria are based on current research and complex analyses, others rest on past product performance and expert opinion, and they have remained unchanged because there is no overt evidence that change may be warranted.

Improving health and fatality statistics suggest that this system has delivered increasingly safe facilities, but we do not really know the levels of overall risk to which facility users are routinely exposed, or the levels of safety that might be achieved through more balanced effort. Facility risks stem primarily from rare events, such as earthquakes and fires, or from slowly accumulating effects of exposure to hazardous conditions. Public awareness of such hazards and how to respond in hazardous situations contribute to reduction of loss. Nevertheless, experience suggests that unnecessary costs are imposed to guard against some hazards while others are relatively neglected.

The evolving discipline of risk analysis, as applied to engineering issues, is an outgrowth primarily of the nuclear power and defense industries. This discipline offers the next step in improving facility safety and the safeguarding of property values, a promising means for facility professionals and the nation to improve the overall safety of its facilities—both new and existing—without imposing unacceptable costs. Federal agencies and the private sector should work to adopt risk analysis procedures more broadly in planning, design, construction, operations, and maintenance.

Risk analysis is, above all, a very effective way of thinking about how public health, safety, the beneficial uses of a facility, and property values may be protected from failures of facilities to perform as anticipated. As a set of tools and procedures used to characterize, typically in a quantitative manner, the threats to safety posed by specific hazards, risk analysis procedures may be applied at various levels of sophistication and detail to support effective risk management.

Risk management is a much broader activity than technical risk analysis alone. Risk analysis provides improved information for risk managers to use in exploring their options and making decisions to improve safety and protect property values. Those who plan, design, and manage facilities seek to manage risk within the context of the anticipated behavior and preferences of people who occupy and own buildings and the costs of practical actions designed to avoid hazards or reduce the consequences of hazardous occurrences. Their ability to effectively manage risk is often limited by a lack of adequate data and effective analysis. While new computer-based analysis tools are emerging, greater effort is needed to bring techniques of risk analysis more quickly out of the universities and research laboratories and into general application.

There are now no generally accepted practices of facility risk analysis and management. The nuclear power industry and structural community have adopted probabilistic analysis procedures that represent limited applications of risk analysis, but broader adoption of a common terminology and broader application of such methods to multiple hazards in and around buildings are needed. Lacking a common framework for discussion and analysis of safety, the public and government officials are often poorly prepared to deal effectively with issues related to events that have small probabilities of occurrence and the potential for severe consequences. Development and broad application of risk analysis procedures will help facility professionals, the policy makers responsible for assuring safety, and the people and property owners exposed to risk to understanding more clearly the nature of those risks and to determine what levels of risk are socially and economically tolerable. Such understanding and clear communication about acceptable levels of risk will then help to assure that safety is achieved and property values are protected as efficiently as possible.

The following actions should be taken to enhance safety through greater use of risk analysis:

1. **Cost-effective procedures of risk analysis should be appropriately applied in design, construction, operations, and maintenance,** to encourage greater forethought and better allocation of resources in managing risk.

2. **Efforts to apply risk analysis procedures should initially be focused in several specific areas:** (a) design and operations of individual high hazard facilities, (b) quality control and code enforcement in construction, (c) facility operations and management activities, (d) facility maintenance, (e) retrofit strategies for dealing with newly identified hazards, (f) strategies for emergency response to hazardous events such as fire, severe storms, or landslides, (g) development and revision of building codes and design guide criteria, (h) evaluation and certification of new materials or technologies, and (i) public discussion and decision-making about standards, codes, and project approvals.

3. **Federal agencies should adopt a risk-based approach** to establishing their facilities planning and design criteria, construction quality assurance procedures, operating policies, and maintenance practices, particularly in high hazard situations.

4. **The national model code organizations and state and local building codes should accept specific applications of risk analysis** procedures for assessing safety of facilities where large numbers of people, especially severe hazards, unusual design or operating characteristics, or unusually high strategic or economic value may lead to unusually high risk.

5. **Facility managers and public officials responsible for regulating building occupancy should adopt risk analysis principles and procedures** to

ensure that (a) operating and maintenance practices or facility renewal activities do not contribute to increasing risk, and (b) needed actions are taken to control newly recognized risks.

6. **Federal agencies should fund additional research** to characterize risk in facilities and to develop more general measures of hazard and risk from their shared experience in facility performance. These agencies should work through the model codes and standards organizations to motivate the private sector to use these measures as well.

7. **The Federal Construction Council should bring together the federal regulatory and construction agencies, the insurance industry, the codes and professional organizations** that deal with building standards, and educational institutions to consider establishing institutional mechanisms for systematic collection and sharing of data on facility failures and hazards incidents, to enable broader application of risk analysis in and around buildings.

8. Government agencies, model codes organizations, building professionals, and others responsible for assuring facility safety should work—through professional education and training and communication with policy makers, facility owners, and users—to **increase general awareness of how application of risk analysis principles and practices can be used to improve safety and protect property values at reasonable costs.**

The benefits to be gained by applying risk analysis to facilities include early identification of design weaknesses; better allocation of resources to achieve balanced reduction of risks; better recognition of the role that human action in design, construction, operation, and maintenance plays in raising or lowering risks; and improved ability to recognize and respond to new hazards or increasing risks. These benefits—for facility owners, occupants, and neighbors—will mean lives and dollars saved.

STUDIES IN MANAGEMENT OF BUILDING TECHNOLOGY

This report is one of a series of products of the Building Research Board's strategic program in Management of Building Technology. An interdisciplinary field of study rather than a recognized discipline, management of technology links engineering, science, and management to plan, develop, and implement technological capabilities to shape and accomplish the strategic and operational objectives of an organization. Observers of the U. S. construction industry have expressed concern that failure to manage technology effectively—at the level of the nation and the individual firm—is a primary factor underlying a perceived risk that U. S. industry is losing its competitive edge in an increasingly global marketplace. These observers argue that action is needed to deal with issues such as liability and societal risk aversion, short-term perspectives, and traditions that divert resources and discourage innovation in both the processes of construction and in facilities. The Building Research Board has undertaken, through this strategic program, to focus discussion and stimulate appropriate response to such issues.

This study was supported as part of the technical program of the Federal Construction Council (FCC). The FCC is a continuing activity of the Building Research Board, which is a unit of the Commission on Engineering and Technical Systems of the National Research Council. The purpose of the FCC is to promote cooperation among federal construction agencies and between such agencies and other elements of the building community in addressing technical issues of mutual concern. The FCC program is supported by 16 federal agencies: the Department of the Air Force, the Department of the Army (2 agencies), the Department of Commerce, the Department of Energy, the Department of the Interior, the Department of the Navy, the Department of State, the General Services Administration, the National Aeronautics and Space Administration, the National Endowment for the Arts, the National Science Foundation, the U.S. Postal Service, the U.S. Public Health Service, the Smithsonian Institution, and the Department of Veterans Affairs.

The Public Facilities Council (PFC) was formed in 1983 to make available to state and local governments, quasi governmental authorities, and others, the forum and services of the BRB and NRC to identify technical problems and research needs facing construction administrators and facilities managers. Sponsors of the PFC currently include a score of state and local governments or interstate entities. Funding and participation are typically drawn from the executive office of the jurisdiction responsible for facilities development and management.

Reports resulting from Building Research Board programs are provided free of charge to sponsoring entities.

For information contact:

Director
Building Research Board
National Research Council
2101 Constitution Avenue, N.W.
Washington, D.C. 20418

CONTENTS

APPENDIXES

USES OF RISK ANALYSIS TO ACHIEVE BALANCED SAFETY IN BUILDING DESIGN AND OPERATIONS

1

INTRODUCTION

Buildings house a variety of human activities, and must do so with a high degree of safety and security. Many of the criteria used in facility design and the standard practices adopted in construction are intended to deal with threats to safety and economic losses posed by such hazards as fire, wind, earthquake, toxic materials, criminal activity, or potential misuse of facilities. Some of these design criteria and construction practices are formally stated in building codes or guidelines used by facility professionals. Others are imposed by law, instilled through professional education, and enforced by the practices of professional organizations and building trades unions.

The influence of these criteria and practices on safety is presumed on the basis of past experience, scientific analysis, and reasoned discussion by those concerned with protecting the public-at-large and the interests of property owners. This concern is frequently shared by the members of professional organizations, trade groups, government bodies, and public interest groups. These groups also share two key problems: a) limited resources, knowledge, and information that reduce their ability to make effective judgments about what should be done to ensure that the nation's facilities are adequately safe, and b) choices that must be made when improvements to safety require increased cost or reduce achievement of other desirable characteristics of the building. Judgements about facility safety must be made within a complex context of the many costs, benefits, goals and objectives of a facility. [See box.]

IS FIRE RISK SERIOUS?

The headlines are invariably bold. Manhattan, 1911, the Triangle Shirt Waist Company: "141 MEN AND GIRLS DIE IN WAIST FACTORY FIRE." Boston, 1942, the Coconut Grove Nightclub: "300 KILLED BY FIRE, SMOKE AND PANIC IN BOSTON CLUB." The Bronx, 1990, Happy Landing Social Club: "87 DIE IN BLAZE AT ILLEGAL CLUB." Such horrible disasters attract national attention and motivate intense reviews of local building code regulations, enforcement procedures, and building design and products characteristics.

The 1974 report of the National Commission on Fire Prevention and Control assessed the nation's fire problem as major: 12,000 lives lost, 300,000 people injured, at least $11.4 billion in direct property losses annually. The United States was found to be the leader by far among industrialized nations in per capita deaths and property loss from fires. Other studies estimated the total fire-related economic burden in the United States to be as much as $36 to $45 billion annually, up to 1.4 percent of our gross national product (GNP).

Yet the Commission noted, "The striking aspect of the Nation's fire problem is the indifference with which Americans confront the subject." Fire experts termed the U.S. residential fire problem "shameful."

Today, U. S. fire deaths have been decreasing for more than two decades, both in total and per capita, and current reports place the number at about 5,000 to 6,000 annually, primarily children and elderly victims. The numbers of multiple fatality fires (with three or more deaths, accounting for about 16 percent of fatalities in 1984) have been decreasing as well, but not their severity.

In the litigious climate of the 1980s, claims raised in such noted hotel fires as the MGM Grand and the Dupont Plaza reach into billions of dollars. Debate continues on such questions as installation of sprinklers and alarms in hotels.

How serious is fire risk? Do we need to work harder to reduce fire risks? There is no easy answer.

References: National Commission on Fire Prevention and Control. 1974. **America Burning**. Washington, D.C. Journal of Code Enforcement. Volume II, Number 2, April, 1990; Committee on Fire Toxicology, National Research Council. 1986. **Fire and Smoke–Understanding the Hazards**. National Academy Press, Washington, D.C.; J. Snell. 1989. **Quantitative Evaluation of Building Fire Safety**. Center for Fire Research, National Institute of Standards and Technology, Gaithersburg, Maryland.

The need to make informed judgments about safety and liability is hardly unique to facilities. Concerns about the safety of new drugs and other medical technologies, food additives, pesticides and other materials that may pose threats to the environment, and nuclear-powered electric power generating plants or other facilities that could fail with possibly catastrophic human, environmental, and economic consequences have, in recent years, motivated development of the principles and practices of **risk analysis.**[1]

Risk analysis is a set of tools and procedures used to characterize — either qualitatively or, more typically, in a quantitative manner — the threats posed by specific hazards. The procedures—focused on identifying potential, hazards and the sequences of events that can lead to losses and the magnitude of possible losses —are typically based on principles of probability theory and statistical analysis, and may involve complex judgements about health, productive work, and the value of human life and property. While these judgements often invite controversy, risk analysis is selectively but increasingly being used in government policy development and regulatory decision making, in the nuclear industry, in food and drug regulation, and in the management of environmental hazards that pose threats to human life and health and property.[2]

Risk analysis is being applied in a limited way to facilities, and these applications have been a subject of debate.[3] Critics argue that risk analysis is too uncertain to be useful for facilities design and management. They also argue that, in any case, using risk analysis implies tolerance for risk and acceptance of lower levels of safety than many people expect in facilities. Proponents, noting that life is uncertain and risk is unavoidable, suggest that risk analysis is a valid and valuable aid to decision-makers seeking to use limited resources to enhance the overall safety and efficiency of facilities.

The potential for greater use of risk analysis—to enhance the overall safety of buildings—motivates this study. More specifically, the question was asked whether federal government agencies should make broader use of risk analysis in developing their facilities design criteria.

---

[1]The terms risk assessment, risk analysis, and risk appraisal are found in the literature of this still young and rapidly evolving field. In this report, these terms are meant to convey similar meaning and intent. "Risk analysis" was selected as the generally preferred term for use in this study. Risk management is a broader term that signifies the active effort to control and reduce the risks faced by an organization or enterprise.

[2]Issues related to this use have been the topic of other NRC committee studies (NRC, 1983 and 1989).

[3]See, for example, Rowe (June, 1987). Nearly a decade ago, a study by the National Institute of Building Sciences (NIBS) recommended that steps be taken to assist the building community to understand, accept, and use risk analysis techniques. (NIBS, 1982)

SOURCE AND SCOPE OF THE STUDY

The sponsors of the Federal Construction Council (FCC)[4] requested the Building Research Board (BRB) to assess the merits and costs—relative to current practices—of broader application of risk analysis in federal facility design. The Committee on Risk Appraisal in the Development of Facilities Design Criteria was appointed to make this assessment and to recommend whether risk analysis techniques can foster efficient risk reduction in facilities.[5] The committee and its individual members reviewed available information, considered presentations made by government officials and professional organizations, and met several times to conduct their assessment. This document reports the committee's conclusions.

The committee focused considerable attention on the extent to which current risk analysis techniques can contribute effectively and at reasonable cost to reducing risks in and around buildings. Having concluded that the potential for such contribution is substantial, the committee then undertook to identify opportunities for applying risk analysis techniques so that they can make greater contributions to balanced safety in the future.

The committee posed questions in three principal areas:

1. What is the nature of "risk" in constructed facilities? Under what conditions does this risk warrant special treatment, not only in design but in operations and maintenance as well? How may recognition be given to the role of human error in subverting the effectiveness of design safeguards?
2. What should be the scope of risk analysis for buildings and other constructed facilities? To what extent should analysis include risk management strategies, as distinct from simply the technical appraisal of levels of risk?
3. In view of the study's sponsorship and anticipated audience, and of the extensive literature and range of current professional activity related to various

---

[4]Sixteen federal government agencies with broad interests in building and facilities research, design, construction, operations, and maintenance sponsor the FCC. These agencies control a major share of the nation's public fixed assets, and have a combined annual construction budget exceeding $7 billion. Some of these agencies already make extensive use of risk analysis, most notably the Department of Energy, which is responsible for a number of high-hazard facilities.

[5]Biographical descriptions of the committee's membership are presented in Appendix A.

aspects of risk analysis, how can the committee's work be most useful to the sponsors and to the facilities design and management professions?

In working to answer such questions, the committee noted the current tendency to try to use legislation and regulation to avoid all risks. Such risk avoidance, in part at least a reaction to perceived inadequacies in public policy, may unnecessarily discourage new technology and lead to actions that have uncertain or poorly understood consequences, and has been cited as a cause of perceived declines in the rate of technological innovation in the U.S. building industry (NRC, 1988). When legislative, regulatory, and judicial decisions have high public visibility, they unavoidably are made within a political context, and turn on public opinion that often is not well informed by technical analysis. Such analysis frequently can help to reduce the level of uncertainty in this decision-making.

The public depends on facility design and management professionals and on government officials to ensure facility safety, and tends to take it for granted that safety is indeed assured. A fire in a high-rise building or collapse of a roof can spur questioning of current safety standards, and lead to precipitous introduction of new building regulations. Some people argue that broad and explicit acknowledgement of risk in buildings might unintentionally call into question the current system of building safety assurance and could lead to loss of public confidence.[6] However, experience in other fields demonstrates that effective communication and active involvement of the public in identifying risks and determining what risks are acceptable are keys to effective use of risk analysis in public sector decision-making. The committee concluded that its charge necessitated some attention to risk communication.

FOCUS AND STRUCTURE OF THE REPORT

This report is addressed to everyone concerned with facility safety, and especially to those designers and constructors, facilities managers, and others responsible for setting policy that influences efforts to achieve greater safety in and around buildings. The committee concluded that risk analysis should be more extensively applied, in the private as well as public sectors, to help improve the cost-effective use of limited resources to enhance overall safety and protect human, environmental, and economic values. This report

[6]In some extreme cases in other fields, such a loss of confidence has been reflected in public outrage and consequent distortions of sound and well-reasoned public policy. (Sandman, 1988)

describes the factors which support this conclusion and presents the committee's recommendations for broader use of risk analysis.

Risk analysis is a technically complex topic and a rapidly evolving technical discipline. This report is neither a thorough review of the state of the art nor a summary of principles and practices of risk analysis.[7] Rather, it is a consideration of major trends, meant to point the way for those who may undertake or use risk analyses.

Chapter 2 discusses the sources of risk in and around buildings, how risks may be managed to achieve safety and security, and why the committee concludes that greater safety and security can be achieved through broader application of risk analysis. Chapter 3 describes a number of barriers to this broader application and to achieving more effective risk management through use of risk analysis. Chapter 4 presents the committee's conclusions and recommendations for overcoming these barriers and enhancing achievable safety and security. Chapter 5 presents the committee's recommendations for broader use of risk analysis, including immediate actions to foster this use. Appendices provide background on principles and procedures of risk analysis and how risk analysis is being used by some government agencies.

Throughout their deliberations, the committee members were mindful that enhancing facility safety and security—within a complex framework of social, economic, and political forces—is a major challenge. Ultimately, safety and security are influenced by a complex of factors: how people perceive risks, how they act on their perceptions, the costs and difficulties of actions to improve safety, and the possible consequences of taking or failing to take action. Facilities designers and owners, and the government agencies and professional organizations that seek to ensure safety and security, all must make decisions that influence specific facilities. These decisions often strike a balance among sometimes differing points of view on risk and safety. By making assumptions about risk factors more explicit and commonly understandable, broader application of the principles and practices of risk analysis can aid these decision-makers. The committee hopes its work will encourage broader use of risk analysis, and thereby enhance safety and security throughout the built environment.

---

[7]Appendix B reviews briefly some of the major principles and practices of risk analysis, developed and used in a variety of technical applications.

2

RISK AND ITS ANALYSIS IN AND AROUND BUILDINGS

In common parlance, risk is the possibility of loss or injury, a probability that some event will occur with serious consequence. Risk is inherent in all human activities, an unavoidable aspect of life, and a concern in all aspects of buildings and other constructed facilities.[8] Fires, extreme weather, and earthquakes, are among the more obvious sources of risk that occupants and owners encounter in and around buildings, and are examples of the large variety of threats to people's safety and investment in property. Building professionals and government authorities have developed extensive design rules and building regulations in an effort to maintain risk at what seems to be reasonable levels.

Risk stems from many sources, and the levels of risk judged to be reasonable may differ from one community to another and from time to time. At a national level, an extensive and aging inventory of existing facilities is a source of concern that risk may be growing. Historic preservation of old structures, in particular, may pose problems when extensive physical changes

[8]Designers and managers may deal with risk only implicitly, through use of safety factors and other design standards. Risk is sometimes inadvertently neglected through adoption of reasonable assumptions (e.g., the chances that a severe storm will occur), or may be intentionally understated in order to circumvent obstacles to a particular project or activity.

and adaptive reuse[9] expose occupants and owners to hidden or unexpected problems. Through aging, any facility may experience progressive growth of risk, due to deterioration of building materials or other natural physical forces. However, the committee judges that as many as 90 percent of structural failures, regardless of structural age, may be attributable to human error at some stage of the facility development and use rather than to catastrophic natural events.

DEFINING RISK

Risk[10] arises because of a specific hazard—an act, event, or phenomenon—posing potential harm to people or activities or things. Fire, earthquake, wind storms, flooding, toxic and allergenic materials, and terrorist attack are examples of hazards associated with buildings and other facilities. The consequences of a hazard are the elements of harm that might result, including numbers of people exposed and severity of harm—e.g., deaths, injuries, dollar value of property damage, activities disrupted, area affected, legal liabilities, and environmental damage.

The idea of risk includes the magnitude of potential consequences of the hazard and the chances that the harm will be realized, that is, the probability of occurrence of the actual event or act, and subsequent loss or injury. Fire risk in a building, for example, may include fires starting from a number of possible sources and various outcomes depending on detection, alarms and fire fighting response when a fire is detected, weather, and the materials of internal furnishings. Each outcome has an estimated probability of occurrence.[11]

---

[9]Adaptive reuse occurs when a building is converted from the use for which it was originally designed to another purpose. In many older cities, warehouses and factory buildings have been converted to housing or shopping malls.

[10]Unless otherwise noted, the definitions used in this report have been adapted from two NRC reports: **Improving Risk Communication** (1989), or **Risk Assessment in the Federal Government: Managing the Process** (1983), National Academy Press, Washington, D.C.

[11]The distinction between hazard and risk is illustrated by the prospect of crossing the Atlantic Ocean in a rowboat rather than an ocean liner: The water hazard is similar for the two vessels, but the risk is very different.

Safety is improved when risk is reduced. Risk is reduced by avoiding the hazard (e.g., avoiding flooding by not building in a flood-prone area), by reducing the chances of loss (e.g., using shatter-proof glazing in windows), or by limiting the likely magnitude of loss (e.g., using smoke detectors to give people more time to escape a fire). However, hazards and facility response to hazards are uncertain, and the assessed risk in any situation reflects the range of uncertainties as well as the levels of hazard and resistance to damage or loss.

Because there is always some risk, safety is not an absolute condition; it can be discussed reasonably only in relative terms. "Unsafe buildings" are those found to be in a condition that is demonstrably dangerous or a hazard to life, health, property, or safety of the public or occupants (BOCA, 1985), according to defined standards of measurement or the judgment of appropriate authorities. Safer facilities expose their owners, occupants, and neighbors to less risk, i.e., fewer or smaller hazards or lower probabilities of occurrence or some combination of these factors.

Building risk and safety depend on where facilities are located, when and how these facilities were constructed, the activities they house, and how they have been operated and maintained. [See box next page.] Some facilities — for example, research hospitals and military installations — may expose their occupants to unusual risks such as infectious diseases or ammunition explosions. Other facilities such as nuclear power installations or toxic waste depots may present unusual risks for people and activities in the vicinity of the facility and over large areas. Risks for handicapped people, children, or people with particular medical conditions may be greater than for other groups.

SOURCES OF RISK

Risk in and around buildings stems from a wide variety of specific hazards. (See Table 1) From time to time, new hazards are identified and become the subject of debate, public policy and regulation. Radon gas, for example, has been recognized as a potential hazard only within the past two decades, while the toxicity of lead has long been known.[12] The hazards and risks of electromagnetic radiation from such sources as video display terminals, microwave ovens, building wiring, and electrical transmission and distribution lines are still being defined. New technology, design details, or construction practices may give rise to new or greater risks.

---

[12]The severity of risks from both radon and lead-based paint and appropriate programs to respond to the risk are subjects of continuing debate in national policy forums.

CODES CUTTING LOSSES

On October 17, 1989, a major earthquake struck the San Francisco Bay area, causing widespread destruction and interrupting baseball's World Series. The nation watched as experts, politicians, rescue workers, and residents dug out of the rubble, remembered prior earthquakes, and reflected on the possible consequences of even stronger events that many experts say are inevitable.

Altogether, the Loma Prieta earthquake was credited with causing some 60 deaths, more than 3,700 injuries, displacement of 12,000 people from their homes, and property damage and other losses totalling at least $6.5 billion.

The damage could have been much worse. Professional reconnaissance teams agreed that most buildings and lifeline structures (e.g., roads, water and power supply systems) performed well, in large measure due to California's efforts to develop, adopt, and enforce effective seismic design and construction practices. Failures were widespread in older buildings constructed before current building codes and design practices became the rule in California.

Observers reflected on the sobering thought that other areas of the United States that may be as seismically hazardous as California have not yet or only recently adopted best available seismic design standards and practices in their local building regulations.

References: Lew, H.S., ed. 1990. **Performance of Structures During the Loma Prieta Earthquake of October 17, 1989**, NIST Special Publication 778, , U. S. Government Printing Office, Washington, D.C. International Masonry Institute. February, 1990. **The Loma Prieta, California, Earthquake of October 17, 1989; Observations Regarding Performance of Masonry Buildings**, Washington, D.C.

MECHANISMS USED TO LIMIT RISK

Various mechanisms are used in design, construction, operations, and maintenance to limit risks in and around buildings, including government regulations, construction and building inspection, and operations by fire departments. Local and state government building codes[13] and design

[13]39 states have adopted some form of statewide code. (NCSBCS, 1987) However, adoption and enforcement of building codes are, for the most part, functions of local government. While these building codes are typically based on one of three principal national model codes, they usually reflect unique local concerns and legislative processes. There are estimated to be more than

Natural Causes		
Snow, Ice	Tornado	Tsunami, wave action
High winds	Flood	Hurricane, typhoon
Earthquake	waterway flooding	Wildfire
motion of superstructure	Landslide	
soil-structure interaction	Volcano	
Man-made Events		
Fire	Environmental Conditions	Nuclear and conventional attack
Explosion	Noise	Accident (slip/fall)
Hazardous substance release	internal	Others (e.g. lightning, vibration)
nuclear	external	
biological	Pathogens	
chemical	asbestos	
Security (crime)	radon	
breaking/entering	toxic materials	
vandalism/malicious mischief	indoor pollutants	
information security		

Table 1. Potential hazards in and around buildings

criteria used by professionals are the most common and comprehensive of these mechanisms. Federal government agencies, although not strictly subject to local building codes, have adopted similar requirements to protect the health and safety of their own workers as well as the public at large. These criteria and codes generally seek to restrict or eliminate specific hazards, and do not typically recognize the principle that risks cannot be completely avoided. Some codes and criteria do reflect at least an implicit recognition of probabilities of occurrence.[14]

Building codes typically address only about 20 percent of the concerns that an owner's design criteria will encompass (Building Research Board, 1989). Some risks not addressed in codes may be limited through application of

10,000 distinct building codes in the United States.

[14]For example, Long Beach and other communities in California have adopted structural standards for controlling earthquake-caused damage to buildings, based on estimated probabilities of earthquake intensity. The ASTM, the national professional organization that develops many of the standards and guidelines used in building codes, has used risk concepts to address problems of asbestos and fire as well as a number of other issues not related to buildings. (J.N. Dezern, 1988)

standard professional practice or explicit decisions by facility owners, and others—such as health-threatening air pollution and flooding along shorelines—are the subject of national regulation.[15] Some risks—such as electromagnetic radiation—have not yet been debated or determined to warrant regulation.

Generally speaking, facility risks in areas subject to particular hazards (such as earthquake, flooding, or criminal activity) will be higher unless specific actions are taken to control the risk. Some risks increase with age of the facility because of normal aging and wear of materials and equipment, unless maintenance efforts, sound management practices, and appropriately timed rehabilitation efforts slow or reverse such normal deterioration. Facility designers and owners sometimes act to limit risk by installing detection devices (e.g., smoke and ionization sensors, water detectors for flooding, pressure sensors, security devices) to give early warning of increased hazard or probability of loss.

Uncertainties in projecting uses, loads, environmental conditions, and performance of equipment in service are among the factors making it difficult to limit risk using codes and design criteria. Such mechanisms may also be inadequate when new hazards arise. Systems for identifying potential problems as they develop and responding on a case-by-case basis may then be warranted. Fire departments and disaster relief programs are principal examples of this mechanism for limiting risk. Monitoring of building loads and structural deflections and regular inspections of key subsystems (e.g., for corrosion and wear) serve also to manage risk. Inspection, testing, and other quality control activities in design, construction, and operation are undertaken to avoid increasing risk due to errors or oversights of the designer, faulty construction practices, or inadequate operation and maintenance programs. Risk management professionals seek to assure that risks that cannot be physically limited are effectively reduced or transferred through such means as insurance, emergency response planning, and damage control.

ACHIEVING GREATER SAFETY

Mechanisms for limiting risk have a cost, and a balance must be struck between levels of risk and the costs incurred by facility owners, users, or the

---

[15]These hazards are addressed by laws and regulations administered, respectively, by the Environmental Protection Agency (EPA) and the Federal Emergency Management Agency (FEMA). Regulations issued by the Department of Housing and Urban Development (HUD) also apply. However, only some buildings are subject to these regulations.

public[16] to reduce risks. While current design practice, building codes and inspection procedures are intended to assure levels of safety that the public finds acceptable, there is no mechanism for comparing risks from diverse sources. Current practice places emphasis on some sources of risk while seeming to neglect others. [See box next page.] The committee concluded that overall safety can be improved through more effective allocations of resources to manage risk, and that risk analysis is a useful tool for helping those who must make these allocations.

Risk and safety levels currently are seldom measured in explicit terms. The levels of risk and safety that the people find acceptable evolve through a process of professional consensus and public debate.

Professional consensus is developed in the forums of professional societies and model codes organizations.[17] The process may be informed by extensive testing and measurements in laboratories and field situations, and by the experience of professionals working in the field. Such organizations as the Underwriters Laboratory, Factory Mutual Research Corporation, and the National Institute of Standards and Technology, as well as many other federal government agencies and university and corporate facilities, play a key role in accumulating the information upon which consensus is based.

For some hazards, the consensus is expressed as measures of performance that a building should exhibit, and the building professional is left to make informed decisions about how to achieve that performance. For example, many structural configurations are permitted for a building, so long as forces anticipated in the structural members (computed according to accepted methods) do not exceed the strength of the materials (as determined by accepted tests). For other hazards, such as fire, building professionals are presumed to require more specific guidance, expressed as explicitly required design features or management procedures. For example, sprinklers may be required in newly constructed hotel rooms, regardless of the facility's materials and design. However, in either case, the level of risk is judged implicitly to be acceptable if the requirements are met.

---

[16]Overly stringent building codes are blamed by some observers for shortages of affordable housing. Facility failures caused by the Loma Prieta earthquake disrupted financial markets and economic activity within a large region. Private insurance and federal disaster relief spread the costs of losses throughout the nation.

[17]There are hundreds of such groups that propose design methods, technical criteria, and standards that may be used in building design, construction, operations, and maintenance. However, only a few of these groups have broad influence on building practice.

TOUGH TRADEOFFS

Early one Sunday morning, an explosion apparently centered in the laundry room ripped through a three-story motel in Hagerstown, MD, starting a fire that killed four people and injured 10 others. Minutes before the blast, according to newspaper accounts, a motel employee had reported smelling gas fumes to the local natural gas supplier. One guest who escaped the blaze was quoted: "That building went up like a matchbox. The building codes need to be looked into."

A spokesperson for the motel's owner, a national chain, asserted that the building's wood-frame design, built to meet "the most stringent building codes in any locale," was not at fault. Company officials believed the blast to be unrelated to a similar explosion and fire some nine months earlier in another of the company's motels in Billings, Montana. A faulty shut-off valve at a gas-powered clothes dryer was implicated in the Montana explosion.

In a separate investigation motivated by a series of residential explosions and fires in the Kansas City area, **The Washington Post** reported that the National Transportation Safety Board, an agency of the U. S. government, concluded that homes using natural gas are inadequately protected from natural gas leaks. The study faulted federal and state regulators for having failed to require gas companies to install a small valve, costing about $15 to $20, that would cut a sudden excess gas flow. However, the study acknowledged that many residential gas customers—not always the gas company—are responsible for maintenance of their supply lines and would have to install the valve.

The Post also cited a study by the Gas Research Institute, an industry body, that estimated a national cost of $8 billion over 50 years to install the valves. That group concluded there are better and less costly ways to protect against gas leaks.

Who pays and who benefits from risk avoidance and hazard reduction? Building owners and designers, code officials, and the public at large must make difficult choices.

Public debate comes into play when professional consensus cannot assure that safety is adequate or seems to have missed the mark. A particular disaster such as a multi-fatality fire or loss of life in an earthquake may motivate the debate. At other times, debate may spring from discovery of a new hazard, such as threats to health posed by asbestos. In either case, demands are made for action aimed to increase safety, and new regulatory practices are typically the means adopted to assure the increase.

A number of studies have attempted to assess risks faced by individuals and groups in modern society, and to compare the risks associated with various hazards. However, there are no commonly applied comprehensive measures of safety or standards of acceptable safety.

Furthermore, people seem to demand much lower levels of risk when they are dealing with particularly feared or unknown consequences. (See Figure 1

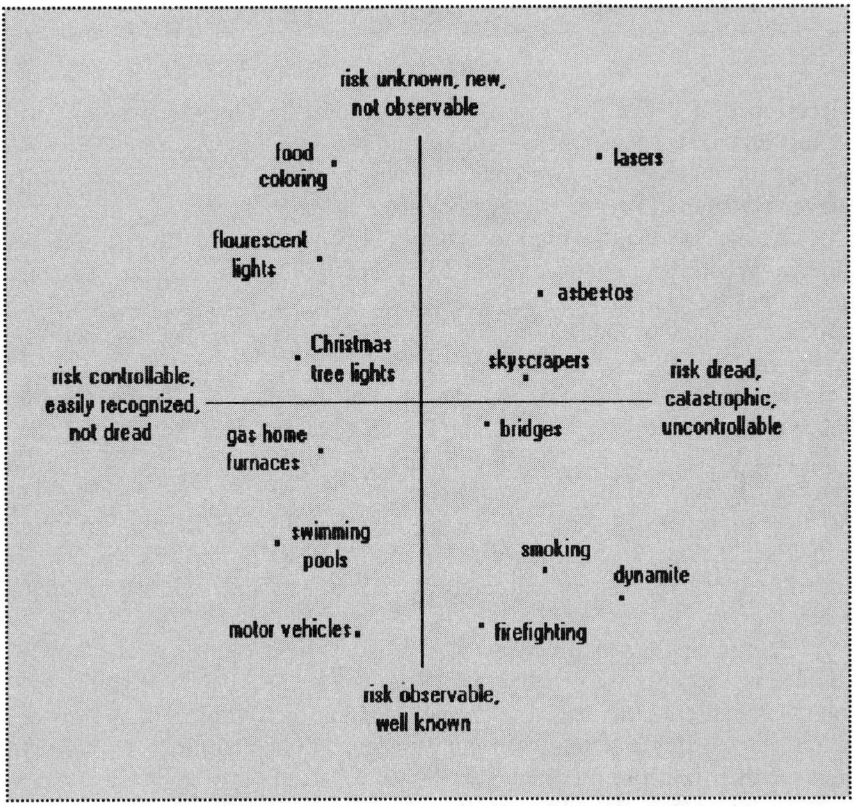

Figure 1. Comparisons of perceived risks from a variety of sources (Source: Slovic, et al., 1985).

for example.) People seem also to accept higher risks when they can choose voluntarily to do so, as compared to situations in which they are exposed involuntarily to risk.[18] (Fischoff, 1984; Kraus and Slovic, 1988) As a result, the levels of risk in and around buildings may differ substantially with respect to various hazards, and may be higher than those associated with hazards from other sources that attract current public concern.

---

[18]For example, highway drivers routinely exceed speed limits and state governments have raised speed limits, despite widespread knowledge that accident severity increases sharply as speeds increase. (TRB, 1984).

WEIGHING RISK AND COSTS

Many risks are treated according to the public attention they attract. Occurrence of a fire that causes loss of several lives in a community will almost invariably lead to reexamination and revision of the fire safety portions of that community's building code. In such a case, the issues of fire safety may overwhelm efforts to manage risk from other sources.

Risk analysis may be used to assure a better balance of effort to achieve safety. Whatever amount is spent to reduce risk in and around buildings, safety will be at its maximum achievable level when these resources are allocated to achieve balanced reduction of risk from all manageable sources, rather than by concentrating on one risk.

Allocation of substantial resources to control risk from one source while other sources are relatively neglected is inefficient and produces safety below optimum levels. Wider application of the logic and procedures of risk analysis can benefit facility design and management. Risk analysis, as an instrument of risk management, can encourage forethought and better informed decisions.[19] Experience in other fields confirms that formal risk assessment contributes to improved overall safety and a better balance of effort to reduce risks.

The principles and practices of risk assessment apply equally well to protection and safety of property as well as people. Probabilistic methods to compute expected net costs or benefits of various courses of action are frequently used by private sector and government decision makers seeking to manage their levels of financial risk. However, the protection of life safety is a primary concern of building professionals and the public, and cannot be addressed in financial terms alone.

---

[19]One example of imbalance in the treatment of risk: Huge sums were being spent in the nuclear industry to avoid accidents caused by a large pipe break leading to a loss of reactor coolant until risk assessment showed that spending these funds on other problems would increase overall public safety. An early federal effort to deal with multiple hazards within a uniform framework also offered little advice on the relative risk posed by these hazards. (Kummer and Sprankle, 1973)

3

USING RISK ANALYSIS TO ENHANCE BUILDING SAFETY AND PROTECT PROPERTY VALUES

Risk analysis, as it has evolved and is currently used, comprises an extensive set of definitions and procedures for characterizing threats to health, safety, mission, or property. Many of these procedures are based on mathematical principles of probability and statistics, while others are simply ways to structure and assist consideration of possible future events and consequences.

Risk analysis has been selectively and successfully applied to the setting of facility design criteria as well as managing risks for specific facilities. For example, designs for nuclear power facilities are subjected to very extensive and detailed risk analysis and, as already noted, building codes that address earthquake hazards are increasingly based on probabilistic analyses.

The entire field of fire safety is undergoing a major advance in risk analysis and management capabilities with the introduction of more realistic and useable computer-based models of fires in buildings. These models will permit explicit analyses of risks to be performed for design and operations planning. While nuclear facilities practices and new fire safety models demonstrate the application of risk analysis procedures, there remains substantial untapped potential for broader application of risk analysis, particularly in federal government programs where one entity effectively retains authority in all phases of a facility's life cycle. However, this is also true in development and enforcement of state and local building codes, and in other phases of managing general building processes.

Risk analysis procedures may be applied at various levels of sophistication and detail to support effective risk management throughout a facility's life cycle, but their effectiveness often is limited by available data and the cost of data collection and analysis. Adequate data are not available to support sophisticated analysis of all facility risks, and more extensive data collection is a pressing need if the fullest benefits of risk analysis procedures are to be realized. Nevertheless, simplified analysis methods, using existing data and professional judgment may be applied more broadly now to enhance risk management.

Communication among the analyst, decision-maker, and the people exposed to risk is an important element of effective risk management. Risk analysis is, above all, a way of assessing hazards and how risk may be managed, a framework for asking the right questions about potential hazards and appropriate responses and searching for good answers. When public and regulatory attention inevitably—and appropriately—focus on elements of risk that are of greatest concern at any given time, risk analysis can inform and facilitate communication, thereby assisting decision-makers who must manage limited resources to achieve the greatest safety and protection.

LOGIC AND PROCESS OF BUILDING RISK ANALYSIS

Risk analysis should start in planning and design. Facility siting decisions influence what hazards need to be considered (e.g., earthquake or coastal zone storms) and the degree of risk (e.g., location relative to potentially unstable slopes or defined flood plain areas).

The body of procedures, criteria, and standards now used in most aspects of facility design represent a distillation and codification of lessons learned from experience and analysis over the course of many years. (Diewald, 1989) The process by which experience enters common practice is evolutionary and often slow, typically involving broad participation of many professionals and industry groups. The levels of risk inherent in any facility are implicitly established in this process, and there is little basis for presuming that risks associated with different hazards are comparably treated.

The traditional approach to design and risk management, based on this accumulated knowledge, reflects an assumption that if the proper procedures are followed and standards are met, unacceptable exposure to hazards will be avoided. If procedures are not followed or standards are violated, exposure is unacceptably high. This traditional logic deals poorly with rare events or new concerns for which there is too little experience to support the evolutionary development of design details or operating procedures that control the hazard.

In contrast to traditional logic, risk analysis methods assume that no prediction can be made with certainty, and that there will always be some chance of exposure to hazards. The process of risk analysis therefore deals explicitly with the possibilities of exposure, the possible consequences of exposure, and the evaluation of these consequences.[20] This logic imposes demands on the analyst and on the institutional setting within which facility safety is managed, demands that may become barriers to improving safety through broader application of risk assessment.

The barriers can be overcome. After some two decades of work, organizations such as the American Institute of Steel Construction (AISC), the American Association of State Highway and Transportation Officials, and the American Institute of Timber Construction (AITC) have begun to introduce reliability-based design codes[21] using the Load and Resistance Factor Design format as an alternative to traditional design methods. The American Concrete Institute (ACI) has used similar formats for some years. The recognized value leading to these changes has been the potential to assure similar structural safety levels in facilities constructed of different materials. The committee seeks to assure similar safety levels relative to the variety of hazards to which people are exposed in and around buildings.

NEED FOR INFORMATION AND DATA

The principal model building codes use two broad parameters to describe building characteristics that determine requirements for safety and health: construction type and building occupancy.[22] The definitions of construction

---

[20]These three stages are sometimes distinguished by such terms as risk identification, risk estimation, and risk evaluation (Rowe, 1977).

[21] Reliability is typically measured as the probability that a structure will not fail under defined service conditions (throughout a service lifetime). The term "design code" is typically used to refer to guidelines and principles published by professional organizations. Such design codes become the basis for contemporary professional practice and may be adopted by responsible government agencies as enforceable building codes.

[22]Construction type generally encompasses materials and structural systems (e.g., reinforced concrete, unreinforced brick, wood frame). Occupancy refers to the building's use—i.e., the activities it houses—and to

types and occupancies now used in most building codes have been developed primarily with regard to the contents of buildings, to address fire hazards and, to a lesser degree, seismic and weather-related hazards. A more complete and generally accepted cataloging of construction types, facility use and occupancies to characterize all sources of facility hazards (including operation and maintenance procedures) is needed. For example, materials used in cleaning, as well as construction, operating electrical loads, and the level of maintenance backlog may influence whether the building's heating, ventilating, and air conditioning (generally referred to as HVAC) and electrical systems should be considered potential sources of fire, explosion, or other hazards. [See box next page.]

Once hazards are specified, formalized risk assessment generally is accomplished by first stating explicitly the possible chains of events that could lead to deaths, injuries, and other losses and, second, estimating the probabilities that these various events will occur within some stated period of time. A variety of formalized procedures have been developed that structure this assessment.[23] Typically these procedures use some form of network diagram to illustrate the relationships among elements of a system, how the system is operated, and external events that may jointly lead to failures. Such analyses can quickly become quite complex and technically sophisticated, and are therefore used primarily when dealing with very complex and highly sensitive facilities (such as nuclear power plants) or as a research tool for exploring policy options (for example, development of fire safety codes).

However, the risk analysis process can be applied in a simplified manner with probabilities estimated by the informed analyst. In general, the value of the assessment ultimately depends on the analyst's foresight and understanding of the technical behavior of the systems being analyzed, and on the availability of data to support estimation of probabilities that hazardous events and consequent losses will occur. In turn, the value of the analysis includes the enhanced understanding of risks and opportunities for their management, as well as the explicit estimates of risk levels.

When there is a well documented history of actual observations of the systems and events of interest, probabilities may be estimated on an actuarial or statistical basis. In the absence of statistical data, simulations or modeling, and expert judgement are sometimes used to understand sequences of events or to estimate the likelihood that certain events will occur or both. Laboratory testing and analogy—a form of simulation—may be used to try

---

the intensity of that use or numbers of people potentially exposed to a hazard (e.g., single-family and multi-family residential, educational buildings, auditoriums).

[23]Appendix B describes some of these procedures.

NEW TECHNOLOGY, NEW RISKS

Small computers and other electronic equipment have become pervasive in the modern office. Designers and manufacturers of these electronic machines have adopted new and highly efficient single-phase power conversion technology as a means to reduce physical size, weight, and cost. However, when large numbers of these machines are operated on a three-phase system of the type commonly used in today's office buildings, transformers may become noisy and overheat, connections in pre-wired office partitions burn out, and building wiring failure may occur. These problems may, in turn, cause fires. Yet all systems seem by conventional measures to be operating well within the limits of accepted ratings for safe service.

The problem stems from use of switch-mode power supplies, sometimes called "switchers" or "switching type," that draw power in from the source in short pulses, rather than continuously. This mode of operation, drawing on the typical three-phase alternating current (AC) supply, results in non-linear loads and high harmonic content—i.e., unanticipated irregularities and fluctuations—in the building's system. When switch-mode equipment is mixed in with other types of equipment such as desk lighting, typewriters, and other electrical machinery commonly found in the office, the problem is not so severe and may go unnoticed. In offices with high concentrations of computers and peripheral devices (e.g., printers, external disk drives, and modems), trouble can occur. The total harmonic current in the neutral wire of the three-phase circuit can theoretically reach 1.73 times the balanced—and normally anticipated—current that would normally occur in any phase.

Detection of the hazard requires fairly sophisticated measurements, and solutions include installation of heavier-duty transformers and wiring. More elegant solutions are possible, such as installation of electrical filters to screen the supply system from the switch-mode harmonics, but such filters are large, costly, and not widely available.

Electrical engineers are coming to recognize that harmonics and non-linear loads will increasingly be problems that must be addressed in design of new offices.

References: Arthur Freund, "Double the Neutral and Derate the Transformer--or Else", **EC&M**, December, 1988, pp 81-85. David Kreiss, "Harmonic Analyzer Helps Solve Power Problems, **EC&M**, March 1989, pp 73-76.

to gain understanding of the expected behavior of new materials or subsystems.

There is no comprehensive data base to support broad assessment of risk in and around buildings. Data that are available cannot easily be used to deal with diverse hazards. Accident and loss statistics form the basis for setting insurance rates, for example. However, the way in which the insurance industry collects and maintains these data typically aggregates data for diverse hazards, records loss in monetary units only, and includes no information on hazard severity. Such data have limited value for characterizing the causal relationships that form the basis of technical risk analysis. The lack of a statistical data base for risk analysis is a major obstacle to broader application of these techniques for improving facility safety.

Investigations of particular facility or component failures are a valuable source of information about causes of failure, and may yield insights that support estimation of probabilities of similar circumstances occurring elsewhere and leading to failure. From the perspective of improving technical understanding of risk and its management, it is unfortunate that such investigations are often conducted within the context of insurance claims and court litigation procedures, so that detailed data are not made generally available for use by researchers. Knowledge gained in these investigations often does enter professional practice and contributes to evolutionary change in design parameters and building regulations.

Often, expert judgement may be the best available basis for estimating probabilities, particularly when new products and techniques or unique facilities are being considered. The public acknowledges that such judgement is a reasonable basis for establishing acceptable risk, and uses professional licensing of architects and engineers to control who is qualified to make this judgment. When failures occur, the courts may be called upon to confirm that reasonable care was exercised in making and acting on this judgment.[24] For complex situations that go beyond the range of normal design and management practices, formalized procedures[25] may be used to synthesize the judgments of groups of experts into a consensus.

BENEFITS AND COSTS OF RISK REDUCTION

The risk analysis produces not only an assessment of overall risk, but also insights into how action can be taken to reduce risk. Certain steps in the

---

[24]For example, collapse of a suspended walkway at the Hyatt Regency Hotel in Kansas City (MO) led to sanctions against the engineers for design errors; severe structural damage in a parking garage in Hartford (CT) was attributed to the owner's failure to carry out effective maintenance.

[25]The "Delphi technique" (also termed the "expert opinion technique"), for example, solicits individuals' judgements on a specific question, compiles the responses, and invites each participant to reconsider his or her earlier judgement in light of the compilation. Repetition of the process generally brings the group toward a group consensus judgement on the question posed. The name, derived from the location of the famous oracle in ancient Greece, reflects the technique's origination as a method for forecasting the future. (Linstone and Turoff, 1975)

chain of events that may lead to loss are typically found to have high probabilities or serious consequences that make them critical to the overall risk. Risk will be reduced if actions can be taken in planning, design, construction, operations, or maintenance to reduce the probability that these events will occur or to control the consequences if they do occur. Analysis of the criticality of facility subsystems or stages in the building process and life cycle indicates where effort to improve overall safety is most likely to have the greatest effect.

Safety of the occupants is a paramount concern of a facility's designers, but is only one of a variety of factors that influence specific design and management decisions. Other aspects of performance, as well as costs of construction and operation, must generally be considered along with safety in the design process. The same balancing of concerns is required in facility operations and maintenance. Achievable safety or allowable risk are then established—in principle—by a comparison of benefits and monetary costs of design decisions and subsequent actions that will influence hazard exposure and consequences.

When particular hazards are covered by the provisions of building codes or owner's design criteria, this comparison of costs and benefits is made at a general level for all buildings covered by these design criteria. Sometimes the comparison has not been explicitly considered, and experience may show that costs and benefits are poorly balanced, i.e., that costs are too high for the apparent improvements in safety or that greater improvements in safety could be achieved at more modest cost. For example, committee members noted that risk analysis showed that huge sums being spent in the nuclear power industry to prevent large-break loss of reactor coolant accidents could be applied to other measures that would substantially enhance overall public safety.

Risk analysis could assist public officials to formulate reasonable responses to occasional disasters that motivate public concern, for example the call for tightening local building codes that frequently follow major fires. Risk analysis could provide similar assistance when new information (for example, scientific evidence that earthquakes are more likely than previously thought) suggests that new actions are required to ensure public safety.

The comparison of benefits and costs may be made for individual facilities and for programs that will lead to the construction of several facilities, as well as for communities as a whole. As in the cases of setting design criteria, increasing experience and new information may lead to a reassessment of whether risks are at acceptable levels. However, the means for responding to new conclusions are restricted to changes in operation and maintenance procedures, and retrofit or reconstruction of the facility. In extreme cases, a facility may be decommissioned and demolished.

Benefit-cost analysis may be used to support these decisions, but great care is required. The tendency of benefit-cost analysis to express all elements of benefits and costs in terms of a single measure—frequently monetary—is always fraught with uncertainty, particularly when efforts are made to place values on human lives and environmental quality. The institutional setting within which risk and safety decisions are made does not utilize economic information alone, and efforts to reduce decisions to strictly economic terms may founder.

INSTITUTIONAL SETTING FOR SAFETY AND RISK MANAGEMENT

Responsibility for safety in and around buildings is distributed among owners, designers, constructors, the insurance industry, regulatory agencies, building occupants, and the public at large and is subject to interpretation and redistribution by the courts. Owners of facilities, especially those that pose particular risks (such as facilities containing hazardous materials), bear the primary responsibility for ensuring their facilities' safety. Architects and engineers (A/Es)—operating as general and specific agents for these owners—help the owners to understand what these risks are and do not bear greater burdens of liability for having done so.[26]

In contrast, the architects and engineers who design other types of facilities may be directly liable for losses. Integrated A/E firms typically pay one to five percent of their gross receipts for adequate insurance, while structural engineers may pay 8 to 10 percent. Firms with good reputations and low loss records may pay less. In addition, there are often limits on the amount of coverage available (currently $15 million is typical), and coverage is subject to a deductible of 1 to 2 percent of the gross damages.[27] Many firms purchase less coverage than the maximum available, or are completely uninsured.

A/Es working in such an environment may hesitate to use risk analysis, which focuses so directly on uncertainties in the building process. A/Es working with federal agencies may find that more knowledgeable staff, less

---

[26]Regardless of who their client is, A/Es will try to inform the client of project risks. One might expect these A/Es generally to avoid possible exaggeration of the risks, out of concern that the owner will decide not to proceed with design development or construction.

[27]The committee was told that these limits are attributable largely to an unwillingness of European reinsurance companies to deal with the litigation system in the United States.

aggressive avoidance of ownership costs, and infrequent use of lawsuits to remedy problems are reflected in reduced exposure to financial liability, and may consequently be more willing to adopt risk-based design procedures.

Because insurance transfers at least a portion of the financial burden of risk away from the facility owner, user, or A/E, the insurance industry could be a principal beneficiary of risk analysis. However, the industry has pursued a loss-based approach to business and has not undertaken broad, systematic study of the technical measures of risk associated with hazards in and around buildings.[28] Risk management professionals employed by private firms and state and local governments focus primarily on actions to reduce financial loss exposure, rather than reduction of technical risk. (PRIMA, 1988)

Local building code administrators and other government regulatory agencies assume some responsibility for risk management by adopting codes and standards to which facilities must conform. The adequacy of the facility with respect to the underlying goal of assuring public safety is presumed if the code and standards are met. The setting of these standards is based largely on a consensus of judgments. If risk, in the technical sense, is reflected at all in these judgements, it is typically in an informal and implicit manner. Furthermore, judgments about conformance to code and underlying safety are in the hands of local government code officials or federal agency design and construction supervisors. Subjective perceptions and assessed levels of risk may differ among these individuals and from one situation to another. It is therefore unclear what the risk levels are,[29] although the absence of widespread losses suggests that the codes and standards, as they have evolved, are delivering relatively safe facilities.

Standards issued by government agencies address some hazards not covered by state and local codes or generally used design criteria. For example, the federal Occupational Safety and Health Administration (OSHA) issues regulations that bear on the design, construction, and operation of facilities. Broad environmental risks related to facility location, design, and operations may fall under the control of the Environmental Protection Agency

---

[28]Large multi-hazard facilities (e.g., chemical manufacturing plants) may be subjected to intense scrutiny to identify ways that loss potential may be reduced without explicit analysis of risk. These facilities are often categorized as "highly protected risks" (HPR) and meet safety standards much more stringent than those normally imposed by insurance companies.

[29]This is not true in those cases, such as many Department of Energy facilities, where explicit risk analysis and risk-based design are used for decision-making.

(EPA) and state or local environmental agencies. Each agency may adopt its own approach as a basis for setting its standards.

The process of environmental impact review prior to facilities construction,[30] conducted with public involvement and documented in an environmental impact statement (EIS), may include particular hazards and risk mitigation actions to avoid losses. However, risk is not necessarily addressed and—some observers suggest—may be avoided because of analysts' concerns that the public will not fully appreciate the inevitability of some risk and may respond negatively to explicit risk assessment.

---

[30]The process initially was required only for projects involving federal actions, but is now required in many state and local laws.

4

OPPORTUNITIES FOR ENHANCING SAFETY AND PROTECTING VALUES

While current lack of data and the distribution of responsibilities for safety among many agencies represent obstacles to effective use of risk analysis, opportunities for improving safety are substantial. Peoples' lives, health, and property can be protected more effectively through broader application of risk analysis.

AREAS OF OPPORTUNITY

Nine phases in the facility life cycle represent particular opportunities for application of risk analysis to enhance achievable safety.

1. Risk analysis can be used as a decision-making tool in the **design and operations of certain facilities**, particularly those which may (a) expose large numbers of people or especially sensitive property to hazards (e.g., sports stadiums, museums, or certain military installations), or (b) be subjected to especially severe or dreaded hazards (e.g., large power plants, medical research laboratories, or facilities housing or employing significant quantities of toxic materials), or (c) involve novel and largely unknown but potentially high risk building technology (e.g., large scale prototype applications of new materials or design methods). Risk analysis procedures are likely to be too

costly and time consuming—in comparison to the possible incremental improvements in risk management—to be usefully applied to all facilities, but in special cases a separate risk analysis may be warranted.

2. Risk analysis can be used by agencies and other facility owners to guide **quality control and code enforcement in construction**. Serious losses associated with human error and faulty practices during construction indicate that facility owners, insurers, and code officials should press for greater attention to the impact on safety of constructable designs, preparation of clear and unambiguous plans and specifications, and construction inspection. Risk analysis can serve as a framework for estimating this impact, guiding quality control and assurance efforts, and assuring that risks are not unnecessarily increased by actions taken during construction.

3. For similar reasons risk analysis can be used to guide **facility operations and management** activities. Personnel responsible for these activities are often unaware of how critically safety depends on a facility's operating systems being used according to the designer's intentions. Budget constraints, time expediency, or simply lack of understanding may motivate changes in occupancy, furnishings, and operating procedures that then sharply increase risk, as outbreaks of legionnaires' disease have demonstrated.[31] Risk analysis may guide operations and management procedures to assure that attention is placed where it matters most to safety. Monitoring of facility use can be an important element of this application, to determine when occupants' activities may be reducing safety.

4. **Facility maintenance**—which should include monitoring of condition and performance—may be the most neglected factor in managing risk. The individuals responsible for the physical aspects of maintenance seldom have full control of resources needed to support their activities, and underfunding is a chronic problem.[32] The collapse of corroded bridges has demonstrated that the consequences of deferral and neglect of maintenance, sometimes slow to become apparent, can accumulate with serious consequence. Risk analysis can be used to characterize the effects of maintenance or its neglect, by drawing valid general conclusions from the individual rare events of facility systems failures.

---

[31]Failure to clean filters in heating, ventilating, and air conditioning systems has been cited as contributing to outbreaks of legionnaires' disease and other respiratory illness among building occupants.

[32]Another BRB committee addressed this issue in their report **Committing to the Costs of Ownership**, National Academy Press, Washington, D.C., 1990.

5. Building professionals can use risk analysis to develop **retrofit strategies for dealing with newly identified hazards**. For example, reported cases abound where the techniques used to remove asbestos increased the risk of exposure, and the committee noted the concern expressed by industry that new regulations to deal with lead-based paints could produce similar results. Risk analysis can help establish rational policies in these situations.

6. Risk analysis can be used to develop effective **strategies for responding to hazardous events** such as fire, severe storms, or landslides. The analyses would encompass maintenance readiness, advance warning, and allocation of resources for immediate response. Risk analysis also provides a framework for reasoned adoption of change in building regulations that may be justified by the new evidence gained from a disaster. Local governments, in particular, might benefit from general guidance, based on risk analysis, regarding when changes in local building codes are or are not warranted.

7. The process for **developing building codes and design guide criteria** in the United States, an already effective means for achieving safety, would be strengthened by greater application of the principles and practices of risk analysis. Probability-based structural design criteria and new work on fire safety (particularly in Canada[33]) warrant greater application.

8. Risk analysis is ideally suited for **evaluating and certifying new materials or technologies**. Systems for approving new building products approval systems are similar in principle to the system used for food and drug regulation. However, there are more than 50 different U.S. product approval systems (FCC, 1990) and few common criteria for judging safety. Risk analysis can provide a common framework for evaluation to support decision-making.

9. Risk analysis can **support public decision making about standards, codes, and projects**. While perceptions of risk vary from person to person and among groups of people, the process of assessing and judging risk seems to enhance the public's ability and willingness to make informed decisions about difficult questions involving risk. More effective communication regarding the risks in and around buildings and the costs of limiting these risks will help public policy-makers to respond appropriately to demands for action following serious losses.

---

[33]The Province of Ontario, for example, has undertaken a review of the cost-effectiveness of all code requirements, and the National Research Council of Canada is developing a risk-cost assessment model for apartment buildings.

OVERCOMING LACK OF DATA FOR RISK ANALYSIS

These nine areas of opportunity are specific instances where greater use of risk analysis has the potential to enhance achievable safety. However, the barriers to this greater use must be overcome for the potential to be realized, and lack of data is the first of these barriers.

One key reason for lack of data is simply the rarity of facility failures from events such as very strong earthquakes, major fires, and other significant hazard events. When these events do occur, they are too seldom observed and measured in a careful way that supports development of a data base for subsequent use by designers and risk analysts.[34] For only a few types of hazard are data on similar events occurring in different locations consolidated and systematically compared to support a more general analysis of risk in and around buildings.

Data on facilities' characteristics that influence risk are similarly sparse because regular condition assessments of facilities are rarely available to support analysis when failures occur. In addition, performance of new materials and facility systems must be estimated from theoretical analyses, laboratory tests, or limited field observations that may not reflect the range of conditions likely to be encountered in practice.

The insurance industry has considerable data on loss experience, but these data are related primarily to dollar losses and industry rate classes, and are poorly suited to risk analysis as discussed here. The industry has developed technical information on how to reduce fire-related and certain other risks in and around buildings, but depends largely on judgment to interpret the degree to which risk is likely to be reduced. Loss experience, a part of the basis for this judgment, depends on costs and a variety of factors not directly related to technical risks.

In contrast to the focused roles of agencies such as the Nuclear Regulatory Commission, the National Transportation Safety Board or the Food and Drug Administration, there is no national regulatory focus for data gathering and analysis of the broader range of building-related risks. The Environmental Protection Agency, the Federal Emergency Management Agency, and the Occupational Safety and Health Administration are among those agencies having some interest in this area, but these agencies' concerns are not focused on facility risks.

---

[34]Analysis of earthquakes may be the most advanced of the principal environmental hazards. Records of ground motion in major earthquakes are available to researchers for detailed study of patterns of damage in the areas where the earthquake occurred. (NRC, 1988)

Some ten years ago, an attempt was made to establish building risk clearinghouse at the University of Maryland. The Architectural and Engineering Performance Information Center started there was never able to attract adequate support, and is not currently active. The National Research Council of Canada undertook to establish a similar effort that also failed to gain momentum.

Professional associations and model building code organizations may partially provide the focus for data base development. For example, the ANSI A58.1 standard on design loads[35] provides a basis for designing typical or average facilities, but offers little information for designing unusual facilities, or for dealing with safety issues related to low-probability hazards.

Government agencies also might play a role in fostering necessary data collection. For example, the Building Research Establishment in the United Kingdom maintains information on failures in buildings and publishes a series of Defect Action Sheets intended to alert the profession to newly understood defects that may be avoided or mitigated. Such a system could be established in the United States as a federal agency such as the National Institute for Standards and Technology.

Lack of generally available data can make analyzing the safety of an existing facility time-consuming and costly. Architectural and engineering plans for existing facilities, especially older facilities, may not be up-to-date or even available. Obsolete materials and construction methods may be unfamiliar to current designers or risk analysts. Hence, a thorough condition assessment and extensive load testing may be required simply to assess the physical characteristics of the facility.[36] Use of risk analysis may, then, effectively be limited to those facilities where the financial investment and numbers of people exposed to the hazard warrant the cost of data collection.

On the other hand, agencies such as the Department of Energy, the Federal Emergency Management Administration (FEMA) and the U.S. Army Corps of Engineers have developed procedures for assessing flood and earthquake hazards at particular locations, and for shaping design decisions

---

[35]American National Standards Institute. 1982. American National Standard Minimum Design Loads for Buildings and Other Structures. New York.

[36]Committee members estimated that current costs for risk analysis in an existing building might typically be in the following range: $30,000 for data collection and load testing for assessment of structural safety in reuse of an older building, and $5,000 for a largely judgement-based analysis of fire hazard in a moderate-sized office building. However, many factors will influence actual costs in any particular situation.

that influence subsequent risks at these locations. The General Services Administration (GSA) has undertaken training programs to inform their staff about the procedures of risk analysis, and principles of risk assessment are reflected in the agency's safety and environmental management programs (Events Analysis, Inc, 1989; GSA, 1988). Such currently active data collection programs as the National Fire Incidence Reporting System, activities administered by the Consumer Product Safety Commission, the National Safety Council, and the Occupational Safety and Health Administration, and loss reporting activities of agencies such as the Navy, Department of Energy, and General Services Administration could be combined or expanded to support comprehensive risk analysis.

The growing experience with such programs may support the research needed to develop more general measures of hazard and risk. Federal agencies should share their own experience in this area, and should fund research and encourage the private sector to use this experience as a basis for developing the measures needed for more general risk analysis.

MOBILIZING FOR RISK ANALYSIS

Even with adequate data, changes are needed in the nation's system for managing risk before the resources of government and industry can be effectively mobilized to use risk analysis. Technical understanding in the building industry has achieved high levels, but facility design and operations still involve the judgement of trained professionals. Furthermore, any complete set of criteria and standards to deal with all risks at all levels would be too costly to apply. However, federal regulatory and construction agencies acting in cooperation with the insurance industry can capture benefits of applying risk analysis, and should join to foster the institutional focus needed for progress. The Federal Construction Council should continue this effort by bringing together these federal agencies, the insurance industry, and the national building codes and professional organizations that deal with building standards, to focus their combined attention on how to implement broader application of risk analysis.

In addition, building professionals should be prepared in the future to use risk data and make judgements about relative safety. Preparation begins with professional education, and the committee recommends that principles and practices of risk analysis be made part of the training received by architects, engineers, and code officials. Texts already have been developed that could be used to introduce information on risk analysis into both university courses and continuing education programs.

COMMUNICATING WITH THE PUBLIC

The levels of safety that can be achieved in practice depend on the public's willingness to allocate resources and undertake the actions required to manage risk in and around buildings. The aim of building designers and managers should always be to make facilities as safe as possible within the constraints of available resources, current technology, and the public's willingness to act.[37]

Current public attitudes toward avoiding risk make it essential that the cost of being wrong—including legal liability as well as physical consequences of underestimating hazard severity or likelihood—should be explicitly considered. The potential for intense public debate may deter government decision-makers from addressing these costs within their usual administrative and political forums. A formal procedure to raise this issue may be warranted for major projects involving high potential hazard.

Experiences from three decades of environmental impact review, show that the manner in which risk analysis is performed in many cases may be as important as the result. People must develop confidence in the process that leads to a decision when they are unable to judge for themselves the quality of the information used in the process. The reputation and professional standing of people involved in the process becomes very important to building that confidence.

Furthermore, public attitudes are influenced, in part, by our increasing ability to identify and measure the intensity of hazards that may previously have escaped detection. Newly identified risks that are poorly presented may foster public dismay and hasty responses that can distort public policy and regulatory practice.[38] Taxpayers and voters may lack a broadly understandable baseline against which to judge acceptable risk and achievable safety.

---

[37]Legal liability sometimes makes policy makers and facilities managers reluctant even to consider that some level of risk is unavoidable and acceptable. Because there is little basis in everyday experience for assessing the significance of very low probabilities, discussion of achievable safety is likely to be controversial and difficult to conduct in a public forum. Sound risk analysis and thoughtful risk communication programs, however, can help to overcome this difficulty.

[38]The opposite effect may also occur. The EPA estimates radon gas to be a significant source of cancer risk in many parts of the country, yet has been unable to mobilize broad public support for regulation.

Comparisons of estimated risk in different contexts (for example, risk of death due to highway accidents versus cancer attributed to a particular drug) are often misleading. Care must be taken to avoid letting illustrative examples become standards for judgement. Nevertheless, "code equivalency"—the levels of risk implicitly accepted for those hazards addressed in building codes—may be a useful baseline for judging levels of risk for other hazards in and around buildings. The same baseline may be useful to federal agencies not subject to local codes but seeking to determine if their design criteria are maintaining risks at acceptable levels. A risk analysis of model codes could produce estimates of what these baseline levels of risk are.

Analysts and decision-makers involved in risk analysis for a particular action (i.e., construction or reuse of a particular facility, or adoption of a new regulatory policy) must recognize that public debate may initially be stimulated as possible hazard events and outcomes are identified. Interested people will seek to assure a common understanding of probabilities and the relative desirability of these possible consequences.

The public participates, in principle, in the current process of establishing building codes and the institutional arrangements that support the code process, but this participation is not consistently effective. The work of the principal national model codes organizations is generally open to public involvement, but building code officials and building products manufacturers are currently the principal participants in those forums, and code changes often are proposed by trade associations and may have economic motivations. Similarly, adoption of official local building codes typically entails public hearings, but the general public often is poorly prepared to deal with the technical issues of facility design and construction.

Nevertheless, there seems generally to be a high level of public confidence in the ability of the design professions and the regulatory process to control risk. Applications of risk analysis should meet highest professional standards to avoid threatening this public confidence.

Experience in other fields illustrates that good public communication is essential. Serious concerns about the siting of hazardous and municipal waste disposal facilities are heightened by people's perception that they are being given incomplete information and being excluded from the decision-making process. A sound risk communication program can help to avoid such exaggeration (NRC, 1989).

5

FINDINGS AND CONCLUSIONS

The benefits of enhanced facility safety warrant concerted efforts to develop the data bases, analysis methods, and institutional incentives needed to foster broader use of risk analysis. Risk analysis procedures foster a useful way of thinking about how mission effectiveness, public health, safety and monetary investments may be threatened by facilities' failures to perform as anticipated. Risk analysis also can improve the allocations of resources to achieve more balanced reductions of risks at all stages of the building process, from early planning through continuing operations and maintenance.

BUILDING SAFETY AND RISK

The traditional body of standards and design criteria embodied in building codes and other documents endorsed by governments or professional organizations bear the primary burden of protecting the health and safety of people in and around buildings. These standards and criteria have, for the most part, been developed and refined through years of experience, and greater attention is sometimes focused more on some hazards than on others. There is no means for judging the overall risk to which facility users are routinely exposed, or the levels of safety that might be achieved through a more balanced effort.

Formal risk analysis, in contrast, seeks explicitly and within a common framework to assess threats springing from a variety of hazards. Engineering risk analysis principles and procedures have evolved primarily within the nuclear power and defense industries to assist decision-makers faced with difficult choices involving rare events with serious consequences. These principles and procedures may be applied to help public officials, design professionals, and other decision-makers responsible for managing resources to achieve facility safety as well.

The committee noted that the nuclear industry and structural community have adopted probabilistic or risk-based methods for facilities design and management (PRA and LRDF, respectively, as noted in Chapter 3 and Appendix B), and perhaps other terms will be forthcoming to describe applications in other limited fields. The committee encourages broader adoption of a common terminology and broader application of such methods to multiple hazards in and around buildings.

AN ANALYSIS PROCESS

The committee highlighted nine specific areas, discussed in Chapter 4, in which risk analysis procedures may be applied at various levels of sophistication and detail to support more effective risk management: (1) design and operations of individual high risk facilities, (2) quality control and enforcement of codes or design criteria during construction, (3) facility operations and management activities, including monitoring of facility use, (4) facility maintenance, (5) retrofit strategies for dealing with newly identified hazards, (6) strategy for responding to hazardous events such as fire, severe storms, earthquakes, landslides, and flooding, (7) refinement of building codes and design guide criteria, (8) evaluation and certification of new materials or technologies, and (9) public decision-making about standards, codes, and project approvals.

A five-part process should be followed to apply risk analysis in any of these areas:

1. Identify the elements of facilities' design, construction, operation, maintenance, environment, and use that are sources of risk.
2. Characterize these hazards in terms of events, and possible outcomes and consequences.
3. Estimate—to the extent practical—the probabilities of occurrence of these hazards, again considering events, outcomes and consequences.
4. Collect the data required to support these estimates of probability and descriptions of hazards.

5. Involve the public through effective communication and appropriate participation in decisions about risk reduction and levels of achievable safety.

The individual steps are not strictly sequential and, to some extent, all parts of the process proceed in parallel. Public involvement will help to define the sources of risk and the nature of these hazards, particularly with regard to the operations and maintenance actions and the role of human error.

Estimation of probabilities and collection of data are also related. Preliminary estimates of probabilities may be used to determine what data will be most effective in sharpening subsequent estimates and resulting conclusions. The uncertainty of final conclusions can be progressively reduced (although never eliminated) until the marginal improvements no longer warrant added costs for data collection and analysis.[39]

Federal agencies that have both responsibility and authority to take action to ensure efficient use of the public's resources should adopt this risk-based approach to establishing their facilities planning and design criteria, procedures for quality assurance during construction, operating policies, and maintenance practices. These agencies' actions will then also demonstrate leadership to encourage private codes and standards organizations to work more rapidly toward balanced efforts for maximum achievable safety.

BROADER USE OF RISK ANALYSIS

Broad use of risk analysis is limited by available data and the cost of data collection. However, simplified analysis methods and existing data can support explicit risk analysis, based primarily on professional judgment, that can enhance overall risk management effectiveness. Government agencies and the private sector should implement the following recommendations:

- **Government agencies, model code organizations, building professionals, and others responsible for ensuring facility safety should work to increase their own and the public's awareness of how risk analysis principles and practices can improve safety at reasonable costs.** Professional education and training in risk analysis should be important elements of this effort, but informing facility owners and users is needed as well to assure that a balanced approach to all aspects of risk management is maintained. Engineers, architects, government officials and policy-makers, model building code

---

[39]This process, widely used in business, is termed Bayesian decision theory. It is named for Bayes' mathematical theorem of conditional probabilities, upon which the procedure is based.

organizations, professional societies and industry groups, universities, and the news media all have important roles to play.

An especially perplexing challenge is finding means to facilitate public dialogue about acceptable levels of risk. Risks can be reduced to very low levels, but cannot be completely eliminated, and those risks associated with particularly dreaded consequences often generate highly emotional debate in public forums. Building professionals and public policy officials should work to inform the public on matters of facility risk, to identify public concerns, and to incorporate these concerns into balanced, cost-effective strategies for risk management.

- **To increase safety and reduce costs, risk analysis should be incorporated selectively into federal agency facility design criteria and state and local building codes.** Codes and formally stated design criteria will remain the primary means for assuring that safety and public health are protected in facilities. Agencies and private groups should work to develop risk-based standards for these documents. While thorough risk analysis is not practical for all facilities, specific application of risk analysis procedures should be required for facilities where large numbers of people, especially severe hazards, or unusual design or operating characteristics may lead to unusually high risk. Unusually high economic or strategic value of the facility or its contents would also warrant explicit risk analysis.

- **Facility managers and responsible public officials should adopt risk analysis principles and procedures to ensure that a) operating and maintenance practices or facility renewal activities do not contribute to increasing risk, and b) needed actions are taken to control newly recognized risks.** Decisions made during planning and design to control risk rely on actions assumed to be taken in subsequent construction, operation, and maintenance of the facility. Sometimes these actions are neglected (e.g., deferral of maintenance), operating conditions change (e.g., new uses of a facility), or new information (e.g., discovery of a new health hazard) indicates that unforeseen action is needed to keep risk at an acceptably low level. Achieving consistently high safety while avoiding unnecessary costs requires attention to risk throughout the service life of the facility.

- **Mechanisms should be established to foster systematic collection of the data required for comprehensive analysis of facility risk.** Government agencies, professional organizations, and the insurance industry are potential beneficiaries of increased use of risk analysis for facilities, and could jointly maintain these data. Investigations of significant facility failures should be made systematically within a common framework, indexed and analyzed to be available to code development bodies, design professionals, and the public. Establishment of an agency or private organization (perhaps similar to the National Transportation Safety Board) may be the most effective way to assure development of a comprehensive data base for facility risk analysis, but individual agencies can start the process by pooling data within common reporting formats[40].

- **Research and development efforts should be accelerated to characterize risks associated with all phases of a facility's service life, and to find effective ways of communicating about this risk to support realistic public judgements about appropriate costs of risk management.** Facility risk is influenced by actions taken in planning, design, construction, operation, maintenance, and renovation or retrofitting of obsolete systems. Work is needed to develop analysis techniques and monitoring systems that support continuing assessment of risk at all these stages. Government agencies, as a group the nation's largest builder and custodian of the public's built assets, should take leadership, working closely with universities and industry.

Early action should be taken on these recommendations. The benefits to be gained include better allocation of resources to achieve balanced reduction of risks; better recognition of the role that human action plays in raising or lowering risks through design, construction, operation, and maintenance; and improved ability to recognize and respond to new hazards or increasing risk. Taken together with the committee's broader recommendations presented in preceding chapters, the benefits can be realized by the industry as a whole and, consequently, by the nation. For facility owners, occupants, and neighbors these benefits of enhanced safety and protection through broader use of risk analysis will result in lives and dollars saved.

---

[40]Studies sponsored by the FCC have pointed the way toward development of integrated data bases for buildings that could include performance data to support risk analysis.

REFERENCES AND SELECTED BIBLIOGRAPHY ON DATA AND METHODS FOR RISK ANALYSIS OF CONSTRUCTED FACILITIES

CITED REFERENCES

Building Research Board, Committee on Advanced Maintenance Concepts for Buildings. 1990. Committing to the Cost of Ownership—Maintenance and Repair of Public Buildings. Washington, D.C. National Academy Press.

Building Research Board. 1989. Use of Building Codes in Federal Agency Construction. Washington, D.C.: National Academy Press.

Dezern, J. N. 1988. Risk Assessment and ASTM: Reasons for and against ASTM's involvement. ASTM Standardization News, Feb. 1988:42-55.

Diewald, W. 1989. Risk Analysis in Public Works Facilities Planning, Design, and Construction. Risk Analysis and Management of Natural and Man-Made Hazards, Y. Y. Haimes and E. Z. Stakhiv, eds. Proceedings of the third conference sponsored by the Engineering Foundation in 1987, American Society of Civil Engineers, New York.

Events Analysis, Inc. 1989. Risk Analysis/Assessment for Safety, Fire Protection, Industrial Hygiene, and Environmental Management Personnel: Student Manual. Published for use by the General Services Administration (August 1989).

Federal Construction Council. 1990. Building Product Approval Systems—Summary of a Conference. Report Number 100. Washington, D.C.: National Academy Press.

Freund, A. 1988. Double the Neutral and Derate the Transformer-or Else. Pp.81-85 in EC&M (December 1988).

Fischoff, B. 1984. Setting Standards: A Systematic Approach to Managing Public Health and Safety Risks. Management Science 30(7):823-843.

General Services Administration (GSA). 1988. Safety and Environmental Management Program handbook. PBS P 5900.2C(August 2, 1988).

Green, M. and A. L. Watson. 1989. Evaluating Buildings in Seismic Zones. Building Standards (July-August, 1989):4-8.

Hammer, W. 1972. Handbook of Systems and Products Safety. Englewood Cliffs, New Jersey: Printice Hall.

Head, G. L., and S. Horn II. 1985. Essentials of the Risk Management Process. Malvern, Pennsylvania: Insurance Institute of America.

Henley, E. and H. Kumamoto. 1981. Reliability Engineering and Risk Assessment. Englewood Cliffs, New Jersey: Printice Hall.

International Masonry Institute. 1990. The Loma Prieta, California Earthquake of October 17, 1989; Observations Regarding Performance of Masonry Buildings. Washington, D.C.

Journal of Code Enforcement. 1990. II(2).

Kaplan, S. and B. J. Garrick. 1981. On the Quantitative Definition of Risk. Risk Analysis 1(1):11-27.

Kraus, N. N., and P. Slovic. 1988. Taxonomic Analysis of Perceived Risk: Modeling Individual and Group Perceptions Within Homogeneous Hazard Domains. Risk Analysis 8 (3):435-455.

Kreis, D. 1989. Harmonic Analyzer Helps Solve Power Problems. Pp. 73-76 in EC&M (March 1989).

Kummer, R. E., and R. B. Sprankle, editors. 1973. Multi-Protection Design. Defense Civil Preparedness Agency, Washington, D.C., TR-20, Volume 6, December 1973.

Lew, H.S., ed. 1990. Performance of Structures During the Loma Prieta Earthquake of October 17, 1989. NIST Special Publication 778. Washington, D.C.: U.S. Government Printing Office.

Linstone, H. A., and M. Turoff. 1975. The Delphi Method: Techniques and Applications. Addison Wesley.

National Commission on Fire Prevention and Control. 1974. America Burning. Washington, D.C.

National Conference of States on Building Codes and Standards (NCSBCS). 1987. Directory of State Building Codes and Regulations, Herndon, Virginia.

National Institute of Building Sciences. 1982. Development and Preliminary Testing of Benefit-Cost and Risk Analysis Methods, Project Report, NIBS, Washington, D.C., November 1982.

National Research Council. 1988. Building for Tomorrow: Global Enterprise and the U.S. Construction Industry. Washington, D.C.: National Academy Press.

National Research Council, Committee on Fire Toxicology. 1986. Fire Smoke - Understanding the Hazards. Washington, D.C.: National Academy Press.

National Research Council, Committee on the Institutional Means for Assessment of Risks to Public Health. 1983. Risk Assessment in the Federal Government: Managing the Process. Washington, D.C.: National Academy Press.

National Research Council, Committee on Risk Perception and Communication. 1989. Improving Risk Communication. Washington, D.C.: National Academy Press.

National Research Council, Committee on Seismology. 1988. Probabilistic Seismic Hazard Analysis. Washington, D.C.: National Academy Press.

Public Risk Management Association (PRIMA). 1988. Public Risk Management: State of the Profession 1987-88, Washington, D.C.

Rowe, W. D. 1977. An Anatomy of Risk. New York: Wiley & Sons.

Rowe, W. D. 1987. Standardizing Safety: Is Risk Assessment a Viable Tool? Standardization News, ASTM 15(6): 38-41.

Sandman, P. M. 1988. Covering Waste Management Controversies: The Hazard and the Outrage; Environmental Communication Research Program. New Brunswick, New Jersey: Rutgers University.

Snell, J. 1989. Quantitative Evaluation of Building Fire Safety. Center for Fire Research. Gaithersburg, Maryland: National Institute of Standards and Technology.

Slovic, P., B. Fischoff, and S. Lichtenstein. 1985. Characterizing Perceived Risk. Perilous Progress: Technology as Hazard, R. W. Kates, C. Hohenemser, and J. Kasperson, eds., Westview, Boulder, Colorado.

Transportation Research Board. 1984. 55: A Decade of Experience. Special Report 204, National Research Council, Washington, D.C.

Wilson, R. 1979. Analyzing the Daily Risks of Life. Technology Review, February 1979:41-46.

GENERAL REFERENCES

The following selected references typify the sort of information on specific hazards and procedures of risk analysis that are relevant to dealing with risk in and around buildings. No attempt has been made to be exhaustive in this listing. The general references illustrate how this information has been used to improve procedures for design and evaluation of constructed facilities.

Risk, Probability, and Safety for Buildings

American National Standard Minimum Design Loads for Buildings and Other Structures. 1982. ANSI A58.1-1982. American National Standards Institute, New York.

American Nuclear Society and Institute of Electrical and Electronic Engineers. 1983. PRA Procedures Guide. Report NUREG/CR-2300, U.S. Nuclear Regulatory Commission, Washington, D.C.

Ellingwood, B. et al. 1982. Probability Based Load Criteria: Load Factors and Load Combinations. J. Str. Div., ASCE 108(5):978-997.

Galambos, T. V. et al. 1982. Probability Based Load Criteria: Assessment of Current Design Practice. J. Str. Div., ASCE 108(5):959-977.

Okrent, D. 1980. Comment on Societal Risk. Science 208:372-375.

Pugsley, A. 1966. The Safety of Structures. Edward Arnold, London.

Rubin, C., et al. 1986. Summary of Major Natural Disaster Incidents in the U.S. 1965 - 1985. Special Publication 17, Natural Hazards Research and Applications Information Center, George Washington University, Washington, D.C.

Slovic, P. 1987. Perception of Risk. Science 236:280-285.

Starr, C. 1969. Social Benefit versus Technological Risk. Science 165:1232-1238.

Wilson, R., and E. A. C. Crouch. 1987. Risk Assessment and Comparisons: An Introduction. Science 236:267-270.

Occupancy live load

Bryson, J. O., and D. Gross. 1967. Techniques for the Survey and Evaluation of Live Floor Loads and Fire Loads in Modern Office Buildings. Building Science Series No. 16, National Bureau of Standards, Washington, D.C.

Chalk, P. L., and R. B. Corotis. 1980. Probability Models for Design Live Loads. J. Str. Div., ASCE 106(10):2017-2033.

Corotis, R. B., and V. A. Dishi. 1977. Probability Models for Live Load Survey Results. J.Str. Div., ASCE 103(6):1257-1274.

Ellingwood, B., and C. Culver. 1977. Analysis of Live Loads in Office Buildings. J. Str. Div., ASCE 103(8):1551-1560.

Harris, M. E., R. B. Corotis and C. J. Bova. 1981. Area-Dependent Processes for Structural Life Loads. J. Str. Div., ASCE 107(5):857-872.

McGuire, R., and C. A. Cornell. 1974. Live Load Effects in Office Buildings. J. Str. Div., ASCE 100(7):1351-1366.

Peir, J. C., and C. A. Cornell. 1973. Spatial and Temporal Variability of Life Loads. J. Str. Div., ASCE 99(5):903-922.

Snow, ice, rain and temperature

Ellingwood, B., and R. Redfield. 1983. Ground Snow Loads for Structural Design. J. Str. Engr. ASCE 109(4):950-964.

Ellingwood, B., and R. Redfield. 1984. Probability Models for Annual Extreme Water-equivalent Ground Snow. Monthly Weather Review 112(6):1153-1159.

Hershfield, D. 1962. Extreme Rainfall Relationships. J. Hydraulics Div., ASCE 88(6):73-92.

Ho, F. P., and J. T. Riedel. 1980. Seasonal Variation of 10-square mile Probable Maximum Precipitation Estimates - United States East of the 105th Meridian. Hydrometeorological Report No. 53, NUREG/CR-1486, U.S. Nuclear Regulatory Commission, Washington, D.C.

Isyumov, N., and A. Davenport. 1974. A Probabilistic Approach to the Prediction of Snow Loads. Canadian J. Civil Eng. 1(1):28-49.

Nicodemus, M. L., and N. B. Guttman. 1980. Probability Estimates of Temperature Extremes for the Contiguous United States. Report NUREG/CR-1390, U.S. Nuclear Regulatory Commission, Washington, D.C.

O'Rourke, M. P., R. Redfield, and P. Van Bradsky. 1982. Uniform Snow Loads on Structures. J. Str. Div., ASCE 108(12):2781-1798.

O'Rourke, M. P., and Stieffel. 1983. Roof Snow Loads for Structural Design. J. Str. Eng., ASCE 109(7):1527-1537.

Steyaert, L., et al. 1980. Estimating Water-equivalent Snow Depth from Related Meteorological Variables. Report NUREG/CR-1389, U.S. Nuclear Regulatory Commission, Washington, D.C.

Wind, Hurricane and Tornado

Abbey, R. F. 1976. Risk Probabilities Associated with Tornado Wind Speeds. Pp. 177-236 in Proceedings of the Symposium on Tornadoes, Inct. Disaster Research, Texas Tech University, Lubbock, Texas.

Batts, M. et al. 1980. Hurricane Wind Speeds in the United States. Building Science Series 124, National Bureau of Standards, Washington, D.C.

Russell, L. R. 1971. Probability Distributions for Hurricane Effects. J. Waterways, Harbors, and Coastal Eng., Div., ASCE 97(1):139-154.

Twisdale, L. A., and W. L. Dunn. 1983. Probabilistic Analysis of Tornado Wind Risks. J. Str. Eng., ASCE 109(2):468-488.

Wen, Y. K., and S. L. Chu. 1973. Tornado Risks and Design Wind Speed. J. Str. Div., ASCE 99(12):2409-2421.

Earthquake

Bernreuter, C., et al. 1984. Seismic Hazard Characterization of the Eastern United States: Methodology and Interim Results for Ten Sites. Report NUREG/CR-3756, U.S. Nuclear Regulatory Commission, Washington, D.C.

Bernreuter, C., J. B. Savy, and R. W. Mensing. 1987. Seismic Hazard Characterization of the Eastern United States: Comparative Evaluation of the LLNL and EPR Studies. Report NUREG/CR-4885, U.S. Nuclear Regulatory Commission, Washington, D. C.

Cornell, C. A. 1968. Engineering Seismic Risk Analysis. Bull. Seism. Soc. Am. 58(5)1583-1606.

Electric Power Research Institute. 1986. Seismic Hazard Methodology for the Central and Eastern United States. Report NP-4726, Palo Alto, California.

Reiter, L. 1986. Current Trends in the Estimation and Application of Probabilistic Seismic Hazard Analysis in the United States. International Atomic Energy Agency, Division of Nuclear Safety TC/NENS-09, Vol. 1.

Explosions, vehicular impact, and progressive collapse

Burnett, E. F. P. 1975. The Avoidance of Progressive Collapse: Regulatory Approaches to the Problem. Report GCR-75-48, National Bureau of Standards, Washington, D.C.

Ellingwood, B., and E. V. Leyendecker. 1977. Approaches for Design Against Progressive Collapse. J. Str. Div., ASCE 104(3):413-423.

Leyendecker, E. V., and E. F. P. Burnett. 1976. The Incidence of Abnormal Loading in Residential Buildings. Building Science Series No. 89, National Bureau of Standards, Washington, D.C.

Mainstone, R. J. 1974. The Hazards of Explosion, Impact and Other Random Loadings on Tall Buildings. Building Research Establishment Current Paper CP 64-74, Garston, Watford, United Kingdom.

Fire

Burros, R. 1975. Probability of Failure of Buildings from Fire. J. Str. Div., ASCE 101(9):1947-1960.

CIB W14. 1983. A Conceptual Approach Towards a Probability Based Design Guide on Structural Fire Safety. Fire Safety Journal 6(1):1-79.

Culver, C. G. 1976. Survey Results for Fire Loads and Live Loads in Office Buildings. Building Science Series No. 85, National Bureau of Standards, Washington, D. C.

Issen, L. A. 1980. Single-Family Residential Fire and Live Loads Survey. NBSIR 80-2155, National Bureau of Standards, Washington, D.C.

Lie, T. T. 1974. A Probabilistic Approach to Structural Fire Safety. Division of Building Research Technical Paper No. 438, National Research Council of Canada, Ottawa.

Lie, T. T. 1979. Safety Factors for Fire Loads. Canadian J. Civil Engr. 6(4):617-628.

Mehaffey, J. R., and T. Z. Harmathy. 1984. Failure Probabilities of Constructions Designed for Fire Resistance. Fire and Materials 8(2):96-104.

APPENDIX A

BIOGRAPHICAL SKETCHES OF COMMITTEE MEMBERS AND STAFF

BRUCE D. MCDOWELL (*Chairman*), is Director of Government Policy Research at the U.S. Advisory Commission on Intergovernmental Relations. He received his BA (Sociology) from American University, MCP (City Planning) from the Georgia Institute of Technology, and PhD (Public Administration) from American University. Dr. McDowell was a practicing urban and regional planner for 12 years, and is a member of the American Institute of Certified Planners. He also has lectured at many colleges in the U.S. and abroad, including Salzburg and Beijing. He has many published articles and reports and has edited or contributed chapters to eight books.

JOHN C. CANESTRO, PE, is a consultant in the fields of codes, construction, and arbitration. He received his BS degree from the University of California in 1949 and a Certificate in Public Administration from California State University in 1974. Formerly a municipal building official, Mr. Canestro has been heavily involved in developing and applying building code requirements and testing methods for energy conservation and construction in seismic areas.

MICHELLE A. DEPEW, is a consultant and formerly the Building Inspector for the Town of Southeast, Brewster, New York, and a Partner in J. W. Shields and Associates Limited. In 1977 she received a Bachelor of Fine Arts degree from Mary-Hardin Baylor University. She has served as an officer and on technical committees of the New York State Building Officials Conference.

BRUCE ELLINGWOOD, PE, is a Professor and Chairman of Civil Engineering at The Johns Hopkins University. He received his undergraduate and graduate education at the University of Illinois at Urbana-Champaign, receiving the PhD degree in 1972. The author or co-author of numerous papers and reports, Dr. Ellingwood is involved in research in the application of methods of probability and statistics to structural engineering. He has served on numerous national professional society and standards committees, and was responsible for the development of the probability-based load requirements for limit states design that appear in the ANSI A58 Standard on Design Loads for Buildings.

ROBERT WILLIAM FITZGERALD, PE, serves as Professor of Civil Engineering at the Worcester Polytechnic Institute. He received a BS degree in 1953, and MS in Civil Engineering in 1960 from Worcester Polytechnic Institute and a PhD in Structural Engineering from the University of Connecticut in 1969. He has written extensively and carried out research in fire science.

GEORGE L. HEAD, is Vice President of the American Institute for Property & Liability Underwriters and the Insurance Institute of America. He earned a BA in General Business from the University of Washington in 1963, an MA in Applied Economics in 1967, and a PhD in Applied Economics in 1968 from Wharton School, University of Pennsylvania. He is a noted authority on risk and loss control management, and is responsible for the Institutes' educational and professional development functions in these areas.

FREDERICK KRIMGOLD, is Associate Dean for Research and Extension, College of Architecture and Urban Studies, Virginia Polytechnic Institute and State University. He received his BA Arch. from Yale University in 1968, attended the University of California in 1969, the Bartlett School of Architecture, University College, London, in 1970, and received his Doctor of Technology from The Royal Institute of Technology, Stockholm, in 1974. A member of the BRB, Dr. Krimgold is a recognized authority on national hazards as they influence building, and recently participated in NRC missions to Armenia.

ANATOL LONGINOW, is a Structural Engineer with Wiss, Janney, Elstner Associates, Inc. He earned a BS degree from Valparaiso University in 1958, MS from Illinois Institute of Technology in 1966, and PhD from Illinois Institute of Technology in 1980. Dr. Longinow is an internationally recognized researcher and educator whose concentration has been in the areas of structural reliability, structural dynamics, and effects on structures of various internal and external dynamic forces, such as chemical explosions, earthquakes, and wind loads.

MORRIS A. WARD, is Executive Director of the Environmental Health Center, a division of the National Safety Council. He is a noted author and commentator, and has managed organizations and publications related to environmental risk and management issues.

CHRIS WHIPPLE, is Vice President and Director, Western Operations, for Clement International Corporation. Formerly Technical Manager, Risk Analysis Program, Environment Division, Electric Power Research Institute, he received his BS degree from Purdue University in 1970, MS in 1971, and PhD in 1974 from California Institute of Technology. Areas of interest include risk assessment methods, risk management and regulation, and the societal response to technological risk. Dr. Whipple has served on several NRC committees and is a member of the Board on Radioactive Waste Management.

ROBERT G. ZALOSH, PE, is Manager of the Applied Research Department and Assistant Vice President of Factory Mutual Research Corporation. He received his BE in Mechanical Engineering from The Cooper Union in 1965, his MS in Mechanical and Aerospace Sciences from the University of Rochester in 1966, and his PhD in Mechanical Engineering from Northeastern University in 1970. Dr. Zalosh is responsible for managing major research on industrial fire and explosion protection and engineering risk evaluation.

Staff

ANDREW C. LEMER, Ph.D., Director, is an engineer-economist and planner. Formerly Division Vice President with PRC Engineering, Inc., and founder of the MATRIX Group, Inc., he is a member of the American Institute of Certified Planners, the American Society of Civil Engineers, and the Urban Land Institute, and received his S.B., S.M., and Ph.D.degrees in civil engineering from the Massachusetts Institute of Technology. His early work on reliability analysis for highway pavement systems has been cited as a basis for current pavement design methods in the United States.

APPENDIX B

CURRENT RECOGNITION OF RISK AND RISK-BASED DESIGN AT SELECTED FEDERAL AGENCIES

The committee invited liaison representatives to describe the types and sources of risks that agencies face in design and operation of their facilities, how they characterize and assess risk, and how they undertake to manage or limit risks through design. This appendix summarizes their presentations. Not all federal agencies that construct or manage facilities were represented by liaisons to the committee.

TYPES AND SOURCES OF RISKS

Risks of death, injury, and property damage due to fire, extreme winds, earthquakes, and other structural loads were acknowledged by all agencies. These risks are perceived to be generally similar to those faced in all facilities, public and private, but in some cases are especially acute.

Military and some civilian agencies identify certain areas of facilities that house functions, information, or equipment critical to a mission's viability, and judge those areas to be particularly risky. Examples include computer facilities, aircraft valued at several times the cost of their hangers, boiler rooms and telecommunications switching rooms, and hospital operating rooms.

Heat and smoke toxicity were identified as principal concerns from fire. Building collapse and trauma injury are generally recognized concerns regarding facility response to seismic loads; however, it was noted that asphyxiation from dust raised by building collapse, rather than trauma (especially in concrete buildings), was a principal cause of death in the 1989 Armenian earthquake disaster.

Several agencies must deal with risks to security. This risk applies to drug storage in medical facilities as well as in military or intelligence settings.

Several agencies must deal with hazardous materials. The Department of Energy (DOE) considers radiological risk in facility siting. The military agencies design to limit risks to military personnel and the general public in weapons firing and munitions storage areas.

The Army's Construction Engineering Research Labs offered a broader concept of risk with respect to application of new technology in facilities. In this setting, risk may be defined with regard to any failure of a system to perform satisfactorily. However, failure and associated risk for wall coverings, for example, differs from that for structural systems.

Some agencies have encountered problems with older technologies that are now being found to have toxic effects on building occupants. Such toxic materials include asbestos and lead in paint and solder used in some plumbing.

CHARACTERIZING AND ASSESSING RISK

Representatives of several agencies indicated that the term "risk" is not explicitly used with respect to design and management of their facilities, even though they deal with risks of the types described in this study. Those agencies that do use the term "risk" adopt a probabilistic approach to assessment, with reference to a specific defined consequence to be avoided. These consequences may differ from one application to another.

The Navy, for example, sets 10^{-7} as the maximum probability per event of injury to military personnel assigned as observers on bombing or gunnery ranges.

The DOE requires that all its facilities be evaluated for potential risks to the operators, the public and the environment in its safety analysis process. Areas addressed in the safety analysis include, but are not necessarily limited to, the following:

- Form, type, and amount of hazardous materials (nuclear or other) to be stored, handled, or processed
- Principal hazards and risks that can be encountered in facility operation, including potential accidents and predicted consequences of fire, explosion, radiation, toxic exposure, structural failure, wind, flood, earthquake, tornado, operating error, failure of essential operating equipment, and failure of safety systems

- Selected design basis accidents such as DBF, DBW, DBE, DBT, OBA, and DBFL, postulated and quantified, including the rationale for selection
- Principal design, construction, and operating features selected for preventing accidents or reducing risks to acceptable levels, including the safety margins used.

The DOE's nuclear facilities siting requirements allow for the use of deterministic (subjective) analyses and/or probabilistic risk analysis (PRA)[41] in judging the selection of one site over others and for overall risk of operation of the facility. For PRA, events considered are those whose annual probability of occurrence exceeds 10^{-6}. For fire protection the DOE requires that its vital facilities and programs meet or exceed the "improved risk" level thereby limiting damage to an acceptable level. As a part of determining the "improved risk" level, the fire protection design analysis addresses those conditions in a facility where the following conditions occur:

- Large or unusual fire potential exists.
- There are special life-safety hazards.
- Toxic chemicals or biological agents exist.
- The consequences of fire include radioactive contamination of the facility, the site, or the public environment.
- National security is adversely affected by fire.

MANAGEMENT AND LIMITATION OF RISK

A number of agencies depend on model building codes for design criteria to control risk in their facilities. Agencies (such as military) have their own extensive design criteria which tend to be similar to those in model codes but also address special mission-related needs. Risk assessment and management becomes an issue, if at all, only on projects that are particularly complex or crucial to the success of a key mission.

Agencies face problems of management and limiting risk in existing facilities as well as new designs. Agencies such as the Postal Service may lease only a few floors in a multi-floor building and yet must be concerned

[41]PRA procedures are used by utilities and the Nuclear Regulatory Commission for analysis of nuclear power plants. These procedures were developed by the American Nuclear Society and the IEEE, working under contract to the Nuclear Regulatory Commission. More broadly applied, such procedures could fulfill many of the committee's recommendations. Refer to Appendix D.

about risk associated with the entire facility. Other agencies suspect that risks are increasing as facilities age. The Army, for example, noted that 43 percent of their buildings are older than 35 years. Some agencies are concerned that newly perceived risks may require costly retrofit of existing structures, as occurred when the seismic hazard zone designation for the city of Memphis, Tennessee was increased.

In two cases—veterans hospitals and ballistics firing to which civilians might be exposed—agencies have adopted an implicit policy of "risk free" design. However, it was generally acknowledged that facilities maintenance and operating procedures can substantially influence the level of risk.

APPENDIX C
ILLUSTRATIVE RISK ANALYSIS METHODS

A number of formal procedures and specific analysis tools have been developed in fields where risk analysis has been applied. Some of the more commonly used methods, briefly described here, illustrate the logic of risk analysis. All such methods have limitations and appropriate ranges of application.

PRELIMINARY HAZARD ANALYSIS

A preliminary hazard analysis (PHA) is a qualitative study of the hazards, components of the related operational system, and event sequences that could lead to an unwanted incident. Possible consequences of the incident and potential corrective actions normally are included in the study. The PHA uses an inductive forward analysis of starting with the failure event and identifying sequential consequences resulting from the failure.

A PHA is an initial effort to identify potential problem areas and the system components and their interfaces. The study is qualitative and considers larger operational components, rather than detailed interactions.(See Figure C-1) If a PHA is done during the preliminary design stage, awareness of the potential problems can enable the designer to incorporate features to reduce, control, or avoid the risk. If the potential risk remains after design and construction, the PHA can be used to identify needs for management or governmental action to mitigate the potential losses.

Hazardous element	Triggering event 1	Hazardous condition	Triggering event 2	Potential accident	Effect	Corrective measures
1. Strong oxidizer	Alkali metal perchlorate is contaminated with lube oil	Potential to initiate strong redox reaction	Sufficient energy present to initiate reaction	Explosion	Personnel injury; damage to surrounding structures	Keep metal perchlorate at a suitable distance from all possible contaminants
2. Corrosion	Contents of steel tank contaminated with water vapor	Rust forms inside pressure tank	Operating pressure not reduced	Pressure tank rupture	Personal injury; damage to surrounding structures	Use stainless steel pressure tank; locate tank at a suitable distance from equipment and personnel

(Source: Henley and Kumamoto, 1981)

Figure C-1 Examples of Preliminary Hazard Analysis (PHA): 1. Hazardous situation: alkali metal perchlorate is contaminated by a spill of lube oil; 2. Hazardous situation: moisture inside pressurized steel tank.

Although the PHA lacks detail, it is a useful procedure to create awareness of potential hazards and to provide guidance for improving protection for the assets of people, property, and operational continuity. The time to complete a PHA is comparatively short, and the experience of the analyst is important to the quality of the results.

FAILURE MODES AND EFFECTS ANALYSIS

A failure modes and effects analysis (FMEA) describes potential failure modes in a system and identifies the possible effects on the system's performance. The method first identifies the subsystem components and their inter-relationships in detail. Then, every mode of failure for each component is considered, and the effects on system function are identified. A FMEA normally is developed in a qualitative format, although quantitative values for reliability may be incorporated. (See Figure C-2)

The FMEA is an inductive, forward analysis. The procedures can involve a high degree of detail because all modes of failure for each component must be considered. A FMEA requires an intimate knowledge of the system, constraints, environment, and the definition of failure. Mathematical sophistication is not important, and reliability quantification of each component can be time consuming and expensive, particularly when components have more than one failure mode.

A FMEA normally deals with equipment and does not include the human action interface, system interaction, and common cause failures. The documentation of failure modes and conditions that lead to the failure enables the designer to focus on details of enhancing reliability and performance in a systematic manner. However, it is often difficult to determine exactly which failure modes cause a specific adverse effect. A fault tree analysis can complement a FMEA to identify causal details.

CRITICALITY ANALYSIS

A criticality analysis (CA) provides a measure by which the relative importance of system components may be ranked. A CA often is an extension of a FMEA. As with a FMEA, a criticality analysis deals mostly with equipment and does not take into account the human activity interface, system interactions, and common cause failures. Criticality may be expressed in functional categories or in probabilistic terms.

Criticality categories may be established subjectively in terms of functional descriptions or levels of damage, or more quantitatively in terms of a

1. Subsystem______ 2. Dwg. Nr.______ 3. Prepared By______ 4. Date______

Item	Failure Modes	Cause of Failure	Possible Effects	Probability of Occurence	Criticality	Possible Action to Reduce Failure Rate or Effects
Motor case	Rupture	a. Poor workmanship b. Defective materials c. Damage during transportation d. Damage during handling e. Over pressurization	Destruction of marble	0.0006	Critical	Close control of manufacturing processes to ensure that workmanship meets prescribed standards. Rigid quality control of basic materials to defectives. Inspection and pressure testing of completed cases. Provision of suitable packaging to protect motor during transportation.
Propellant	a. cracking b. Voids c. Bond separation	a. Abnormal stresses b. Excessively low temperatures c. Aging effects	Excessive burning rate. Overpressurization; motor case rupture during otherwise normal operation	0.0001	Critical	Carefully controlled production. Storage and operation only within prescribed temperature limits. Suitable formulation to resist effects of aging.
Liner	a. Separation from motor case b. Separation from motor insulation	a. Inadequate cleaning of motor cases after lubrication b. Use of unsuitable bonding material c. Failure to control bonding process properly	Excessive burning rate. Overpressurization. Case rupture during operation	0.0001	Critical	Strict observance of proper cleaning procedures. Strict inspection after cleaning of motor case to ensure that all contaminants have been removed.

(Source: Hammer, 1972)

Figure C-2 Example of Failure Modes and Effects Analysis

frequency of failure. The cost of evaluating component failures can vary substantially.

A CA allows a ranking of the hazards and failure effects to identify relative importance or attention. In addition to probabilistic ordering, criticality ranking can be expressed as a product of probability and expected damage.

EVENT TREES

An event tree starts with an identified failure condition and proceeds with a forward, inductive analysis to show causal relationships. This forward analysis traces all possible sequences of events that together describe all possible outcomes of the failure event. The diagrammatic structure that describes these outcomes in a series of discrete, connected events is an event tree. Each branch represents a possible status (state) of the system.

The sequence of events in this forward analysis enables one to identify a number of scenarios (sequences of events) for the outcomes of a single initiating event. The selection of the events reflects the degree of detail that is desired. Figure C-3 shows a relatively gross description of the possible outcomes to a fire in a three room building having Room 1 as the room of origin. Fire termination may occur by self termination, automatic sprinkler suppression, or manual suppression. One could construct event trees to describe events in even greater detail.

The event tree can be used qualitatively to describe possible outcomes of an event. Outcomes may be grouped in terms of their consequences, as illustrated in Figure C-3. When event probabilities are determined, the probability of the outcomes may be calculated. It is often difficult to determine the probabilities objectively because they are conditional on the occurrence of prior events.

The method's inherently binary nature, with each mode indicating a "fail, not-fail" dichotomy can be overcome but is a shortcoming. Systems often degrade without experiencing sudden failure, but detailed analyses of such failure modes introduce considerable complexity into the event trees. Statistical correlations among events may affect the sequence probabilities and are difficult to evaluate with currently available data.

FAULT TREES

A fault tree is a diagram that traces the causal events that can lead to a system failure backward through deductive logic to determine its roots. The events are organized into a logical framework which uses logic gates to

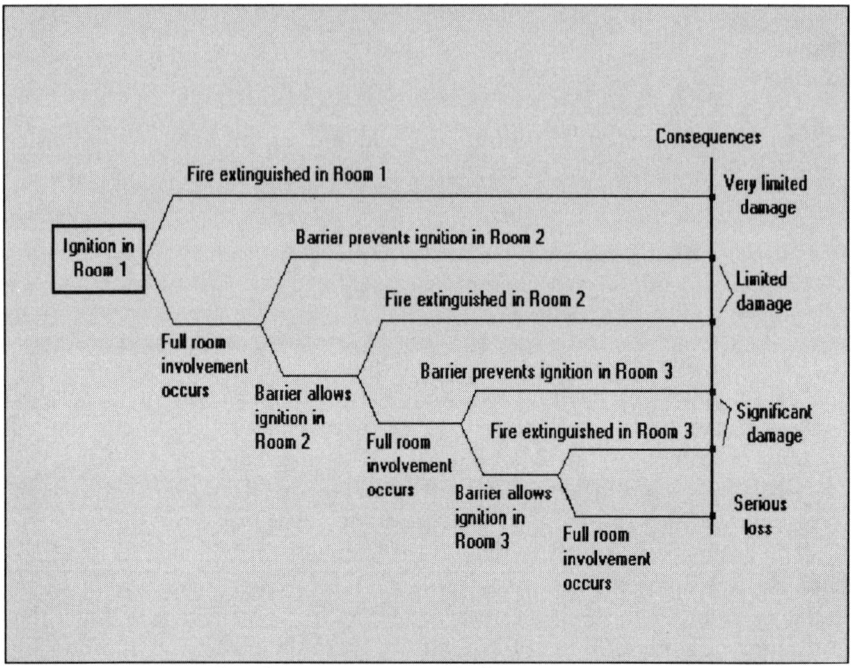

Figure C-3 Example of Event Tree Analysis

identify the causal relationships of the events immediately below the gate. Fault trees can be used for evaluating the events as branches in the event tree in Figure C-3.

Procedures for constructing fault trees are well documented and provide a means by which complex interrelationships can be understood. Construction of a fault tree requires an intimate knowledge of the system being studied, identifying causes, determining exclusiveness, independence, and conditionality of events, and logically describing event interactions require considerable thought and understanding. This process is, however, one of the most valuable parts of the analysis process.

Fault trees are analogous to a photograph in that they depict conditions at an instant in time or as a transition between two consecutive events in an event tree. Fault trees may be used qualitatively to identify events that cause failure. When probabilities are determined for the events, the probability of failure of the main event may be calculated. Time dependent event relationships cannot be represented on fault trees.

Variations of the fault tree analysis may be explored, focusing on actions that will prevent progress toward failure. Success trees and more general network diagrams share with fault trees the basic characteristics of illustrating graphically the chains of cause and consequence that can lead from an initial event (such as an earthquake) to loss of life or safe and acceptable performance.

APPENDIX D

PROBABILISTIC RISK ASSESSMENT[42]

The term Probabilistic Risk Assessment (PRA) refers to the framework and analytical methods widely used for assessing risk in the U.S. nuclear power industries. PRA utilizes event trees and fault trees in a structured analysis process that can be described in five general steps:

1. Identifying various hazards and the frequency of various levels of each hazard (initiating event).
2. Identifying the things that could fail or go wrong and their probability, given the various hazards and hazard levels, including the consequential damage associated with each failure. This step involve the development of all of the scenarios of events that could occur, and their sequence and resultant damage. (The event tree is a tool for this.)
3. Developing the model for each event in the sequences to indicate which of the event's component parts could fail and lead to failure or degradation of the event. (The fault tree is a tool for this.)
4. Quantifying the fault trees and event trees to determine the frequency of each damage state, and thereby determining the ranked order of scenario and the component contributors to the different damage state's frequency.

---

[42]This explanation was provided by one of the peer reviewers of the committee's draft report.

5. Determining the uncertainty in the results by calculating a distribution of damage state frequency, using the distribution of possible hazard frequency and failure probability.